Linear Programming

Narendra Paul Loomba

PROFESSOR AND CHAIRMAN
DEPARTMENT OF MANAGEMENT
BARUCH COLLEGE
THE CITY UNIVERSITY OF NEW YORK

LINEAR PROGRAMMING
A Managerial Perspective

SECOND EDITION

Macmillan Publishing Co., Inc.
NEW YORK

Collier Macmillan Publishers
LONDON

Dedicated to John K. Smart and Surinder Kumar Garg

Macmillan Publishing Co., Inc.
866 Third Avenue, New York, New York 10022

Collier Macmillan Canada, Ltd.

Library of Congress Cataloging in Publication Data

Loomba, Narendra Paul, (date)
  Linear programming.

  Includes index.
  1. Linear programming.   2. Industrial management—Mathmatical models.   I. Title.
HD20.4.L65   1976   658.4'033   75–16343
ISBN 0–02–371630–4

Printing: 1 2 3 4 5 6 7 8   Year: 6 7 8 9 0 1 2

Preface

A number of books have been published on the subject of linear programming. Almost all of these books are mathematical in nature and mechanistic in their approach. In the first edition of this book, the level of the mathematics was very elementary, the role of computers was not discussed, and the target audience was the college undergraduate. In the second edition, we have maintained the same elementary level of mathematics, but the role of computers in realizing the potential of linear programming is described, and the book is directed to both the college student and the business manager. Hence the change in the title to *Linear Programming: A Managerial Perspective.*

The second edition differs from the first edition in three respects. First, the *scope* of the second edition has been extended to include four new chapters. Second, the *focus* of the second edition is managerial, as opposed to technical or mechanical. Third, the material has been revised to include additional useful concepts, to eliminate marginal sections, and to *integrate* thoroughly the twelve chapters.

The purpose of this edition is to provide a managerial perspective on linear programming. A managerial perspective requires that linear programming be discussed in full recognition of the nature and scope of management theory. Accordingly, in Chapter 1 we provide a brief history of management thought and show the synthesis of those vital elements that underlie modern management theory and practice (scientific method, systems approach, and concern for human behavior).

v

Managers work with models; and linear programming is but one of an array of impressive models of management science. However, linear programming is a very versatile model. One way to explain the power of the linear programming model is to discuss its past, present, and potential applications along with all the necessary mathematical formulations and explanations. We did not adopt this approach. No mathematical theorems of linear programming are derived. The focus is placed instead on the managerial utility of linear programming. Linear programming is used as the vehicle by which the construction and utilization of the concept of a model is illustrated.

Managers make decisions. Within the framework of the "managerial perspective," a decision model is discussed in Chapter 2, and a sequence of eight steps is illustrated with reference to a linear programming problem.

The emphasis throughout is on the philosophy and logic of linear programming, rather than on the mathematics and mechanics. Yet, it is very difficult to provide substance to the logic, especially in the area of quantitative models, without the assistance of numerical illustrations. Accordingly, to the extent possible, every concept and method has been illustrated with a numerical example. The level of complexity with which the linear programming model is examined increases slowly as the book progresses. In Chapter 3, a simple problem is examined by way of a *graphical* approach. The same problem is then solved in Chapter 4 by use of an *algebraic* approach.

Chapter 5 presents some of the basic concepts of matrix algebra. These concepts are necessary for an understanding of the remainder of the material in the book.

The *vector* approach, a forerunner of the simplex method, is presented in Chapter 6. The *simplex method* is illustrated in Chapter 7. Also described in Chapter 7 are some special cases of linear programming, including the concept of degeneracy.

Chapter 8 presents the concept of the *dual* problem and explains *sensitivity analysis*. The dual problem and the dual variable have great utility for managers in making economic as well as policy decisions.

The special-structure *transportation model* is described in Chapter 9, and the *assignment model* is presented in Chapter 10. The concept of duality is utilized in the special-structure models of Chapters 10 and 11.

Chapter 11 examines some extensions of linear programming and illustrates *integer programming* by use of the branch-and-bound method. Finally, some important issues and aspects of *application* and *implementation* are discussed in Chapter 12. A computer-produced solution to the same product-mix problem is presented and discussed.

I am grateful to many of my students, friends, and colleagues for their help and advice. Professor Donald Moscato of Iona College reviewed the entire manuscript and made several valuable suggestions. Professor Georghios Sphicas of Baruch College was very generous with his time and advice.

I express my thanks to these members of the Department of Management at Baruch College: Rakesh Gupta, Om Prakash Dhiman, David Cadden, and Robert Stein. Mr. Gupta coordinated and supervised the entire project and helped me improve the book. My thanks also to Professors John Humes, Jack Shapiro, and Lou Stern, all of the Department of Management, Baruch College.

I wish to acknowledge specifically the patience and excellent typing of Betty Kelly. I appreciate her understanding and cooperation.

Last, and most important, I wish to acknowledge the continuing support, understanding, patience, and encouragement that I receive so generously from Mary, my friend and wife.

Narendra Paul Loomba

Scarsdale, New York

Contents

3   LINEAR PROGRAMMING: A GRAPHICAL
   APPROACH   54

4   LINEAR PROGRAMMING: AN ALGEBRAIC
   APPROACH   82

5 BASIC CONCEPTS OF MATRIX ALGEBRA 99

6 THE VECTOR METHOD 137

7 THE SIMPLEX METHOD 150

8 THE DUAL AND SENSITIVITY ANALYSIS 187

9 THE TRANSPORTATION MODEL 207

10   THE ASSIGNMENT MODEL   253

11   INTEGER PROGRAMMING   277

12   APPLICATION AND IMPLEMENTATION   291

1

Management and Linear Programming

1.1 INTRODUCTION

The term *management* has many connotations, implications, and aspects. The nature, role, and importance of management have been discussed in the literature by examining management from various points of view: as a *function, process, art, science, resource, discipline,* and *profession,* and as an *elite* or *class* of people.* We shall view and define management as the *process* of integrating the efforts of a purposeful group or organization. The process concept of management is useful for two reasons. First, it has a time orientation and hence reflects the dynamic nature of management. Second, the process framework emphasizes the decision-making aspects (how choices are made among alternatives) of management. The decision-making orientation is appropriate because linear programming, the subject matter of this book, is one of the most useful techniques for analyzing managerial problems and making managerial decisions.

Linear programming has been employed in solving a broad range of problems in the administration of business, government, industry, education, hospitals, and libraries. As a technique of decision making, it has demonstrated its value in such diverse areas as production, finance, marketing, research and development, and personnel assignment. Determination of optimal product mix, transportation schedules, plant location, machine

* See Levey and Loomba [1973, Chapter 11].

1

assignment, portfolio selection, and allocation of labor and other resources are but a few of the types of problems that can be solved by linear programming.

Linear programming is a method of determining an optimal program of interdependent activities in view of limited resources that are available to the manager during a specified period of time. The term linear connotes the idea that all relationships involved in the problem exhibit proportionality and additivity.* The assumption of linearity is quite useful, and often approximates reality, in such problems as the blending of gasoline, determination of optimal diets, and production planning and scheduling.†

The term program refers to a course of action that covers a specified period of time. A specific production schedule, product mix, stock portfolio, and advertising media mix are examples of programs. Since it is always possible to design a number of different programs, managers attempt to identify those programs which are best or optimal in terms of some measure of effectiveness. A program is optimal if it maximizes or minimizes a measure or criterion of effectiveness such as profit, cost, or sales.

The term programming refers to a systematic procedure by which a particular program or plan of action is designed. Programming consists of a series of instructions and computational rules for solving a problem that can be executed manually or fed into the computer. In Chapter 7 we describe in detail the simplex method, one of several available methods that can systematically solve linear programming problems. The simplex method was developed by the American mathematician George B. Dantzig in 1947.

The term activity refers to any candidate (product, service, project, etc.) that is competing with other candidates (or activities) for limited resources. The allocation of resources among competing activities is one of the most prevalent problems facing managers, and the decision regarding allocation is usually made under the criterion of maximizing profit or minimizing costs. Linear programming is one of the most useful techniques for making allocation decisions.

It should be noted that linear programming is but one of several quantitative models that have been developed during the last several decades. Linear programming, however, occupies a special place in the field of oper-

---

* The meaning of linearity is discussed in Appendix A. Briefly, linearity implies proportionality and additivity. Proportionality refers to the fact that doubling (or tripling) the production of a product will exactly double (or triple) the profit and the required resources. Additivity means that the effect of two different programs of production is the same as that of a joint program involving the same activity levels.

† Whenever the assumption of linearity is not warranted, the problem might be expressed as a nonlinear programming problem. Nonlinear programming deals with those problems in which the objective function and/or constraints are nonlinear. For a simple presentation of some nonlinear programming methods, see Loomba and Turban [1974, Chapter 7].

ations research or management science,* because linear programming has a clearly defined structure, is a powerful tool of analysis, covers a very broad range of applications, and is the product of a philosophy of management characterized by the systems approach, application of the scientific method, interdisciplinary teamwork, and the use of models.†

The main purpose of this book is to familiarize the reader with linear programming, discuss various methods of solving linear programming problems, and use linear programming as the vehicle to illustrate the building and using of management science models. Although we shall discuss and explain the mechanics and rationale of the linear programming model, our orientation will be managerial in nature and our purpose to provide broad *managerial perspective*. After providing the technical details, we shall also address ourselves to the managerial problems of application and implementation (see Chapter 12).

Linear programming is a mathematical model and it is possible to employ different levels of mathematical sophistication to discuss the basic model, its current and potential variants, and its multitude of business applications. We shall, however, restrict our discussion to a level that requires no more than an elementary knowledge of matrix algebra (Chapter 5). The assumptions and structure of the basic linear programming problem and some important properties of linear programming solutions will be forced out of a simple graphical approach presented in Chapter 3. As the book progresses and we move from the simple, but restricted and less powerful, graphical method to the general and more powerful simplex method, we shall have occasion to record several useful *economic* interpretations of certain components of various solution stages of linear programming. These economic interpretations provide policy guidance and can serve as a very useful aid to managerial decision making.

Linear programming, as mentioned earlier, is but one of the several useful decision models that have been developed by management theorists. It is important, therefore, to indicate how linear programming fits into the broad spectrum of management theory and quantitative decision models. Accordingly, after illustrating a practical linear programming problem, we present a brief history of the evolution of modern management. This is followed by a description of management science of which linear programming is but one part. We then briefly explain the meaning of these characteristics of management science: systems approach, application of the scientific method, interdisciplinary teamwork, and the use of models. Finally, we present some selected applications of linear programming.

---

* Operations research (or management science) represents a school of management thought based on the belief that management problems can best be analyzed through the use of quantitative models. See Section 1.4.

† For a description of these characteristics, see Sections 1.5 to 1.8.

1.2 THE PRODUCT-MIX PROBLEM—AN ILLUSTRATIVE EXAMPLE

A linear programming problem arises whenever two or more *candidates* or *activities* are competing for limited resources and when it can be assumed that all relationships within the problem are linear. Let us illustrate a typical linear programming problem by considering a situation in which three products, say *A*, *B*, and *C*, each yielding a specific profit contribution per unit, are to be produced so as to maximize total profit. The production of each product requires the utilization of a specified resource from each of the three departments of a manufacturing unit. The technical specifications of the problem and the production processes are summarized in Table 1.1.

Table 1.1. Process Time Data

	Product			
Department	*A*	*B*	*C*	Capacity Constraint per Time Period
Cutting	10	6	2	2,500
Folding	5	10	5	2,000
Packaging	1	2	2	500
Profit per unit	23	32	18	

What we have described is a *product-mix problem.* The product-mix problem represents a situation where two or more products are competing for limited resources and the manager's task is to determine the optimal product mix.

At this time we make some important assumptions with respect to the information contained in Table 1.1.

1. We assume that, regardless of the level of production, the profit contribution per unit of production remains *constant.*
2. We assume a *fixed* technology. That is, we consider production requirements to be fixed during the planning horizon.*
3. We assume that all decision variables are *continuous*; that is, it is possible to produce products in fractional units.
4. We assume that the manager here is *completely certain* regarding available alternatives (different production programs) and their respective consequences.

---

* By *planning horizon*, we refer to that time period for which the manager is designing the production program.

We shall have occasion to elaborate on these assumptions in later chapters.

On the basis of these assumptions, it is possible to express the data given in Table 1.1 as a linear programming problem. Let us describe the problem and its various components. When we refer to Table 1.1, we note that the products A, B, and C are the activities or competing candidates. The three manufacturing departments represent the resources, and these resources are limited in terms of processing capacities available during a specified time period. Our problem is to design or identify a production program that will maximize profit and yet not demand resources in excess of those that are assumed as given in the problem. In other words, our task is to design or identify the optimal program.

For the problem of Table 1.1, the optimal program is the program that will yield *maximum* profit. If the problem had been of the type in which the manager was mainly interested in costs, the objective would have been the minimization of costs, and the optimal program will result in *minimum* costs. The idea is that in any linear programming problem, the manager must identify a *measurable* objective or criterion of effectiveness. This type of objective can usually be quantified and becomes the *objective function* of the problem. Assumption of linearity implies that it is a *linear objective function*. If we denote the number of units of A, B, and C to be produced by the variables X, Y, and Z, respectively, then, for our problem, the linear objective function is given by

$$23X + 32Y + 18Z \qquad (1.1)^*$$

The objective function, it should be noted, is nothing more than a mathematical expression that describes the manner in which profit accumulates as a function of the number of the three products.

A quick examination of any linear function indicates that its maximization, or minimization, does not make sense without imposing some sort of additional constraints. For example, the maximization of the linear objective function given by (1.1) will result in a value of positive infinity. However, when such a function is subject to certain constraints (e.g., those representing the limited resources), only those values of X, Y, and Z can be considered that will not violate the constraints implied by the problem specifications.

Referring again to Table 1.1, we note that the current capacity of each manufacturing department is fixed for the planning horizon, and hence we have limited resources. The technical specifications of the problem and how

---

* If the quantities of the three products had been denoted by X_1, X_2, and X_3, the objective function would be stated as $23X_1 + 32X_2 + 18X_3$. The linearity of the objective function here is obvious from the fact that each term consists of exactly one variable with an exponent of exactly 1. In large linear programming problems, different variables are usually denoted by $X_1, X_2, X_3, \ldots$. This procedure has the advantage of accommodating as many variables as needed, since the entire real-number system is available for specific identification. At the elementary level of exposition it is convenient to use a system that does not involve subscripts.

they relate to the fixed resource capacities are expressed by a set of *structural constraints*. The assumption of linearity implies that such a set is linear. Since any production program (i.e., values of X, Y, Z) must be such that it cannot demand resources in excess of the given capacities, the *linear structural constraints* for our problem must be expressed as inequalities of the "less than or equal to" type. For our problem, there are three structural constraints:

$$10X + 6Y + 2Z \leqslant 2{,}500$$
$$5X + 10Y + 5Z \leqslant 2{,}000 \qquad (1.2)$$
$$1X + 2Y + 2Z \leqslant 500$$

Although all inequalities in (1.2) are of the "less than or equal to" type, it should be noted here that the linear structural constraints can be any of three types:

Type	Mathematical Symbol
Less than or equal to	$\leqslant$
Greater than or equal to	$\geqslant$
Equality	$=$

A production program, by definition, must be such that a particular activity or candidate is either included in, or excluded from, the production program. This means that negative production, which has no physical counterpart, is not permitted in the solution of a linear programming problem. This obvious fact is made an integral part of the linear programming problem by stating a set of *nonnegativity constraints*. For our problem, the nonnegativity constraints are

$$X \geqslant 0 \qquad Y \geqslant 0 \qquad Z \geqslant 0 \qquad (1.3)$$

We can now summarize the three components of any linear programming problem:*

1. A linear objective function.
2. A set of linear structural constraints.
3. A set of nonnegativity constraints.

* Linear programming is one branch of mathematical programming. Any mathematical programming problem also consists essentially of three components: (1) an objective function that can be linear or nonlinear, (2) a set of structural constraints that can be linear or nonlinear, and (3) a set of constraints on the variables that can be nonnegativity constraints, constraints that specify ranges, or constraints which specify that variables can assume only discrete or integer values. See Loomba and Turban [1974, p. 4]. We present integer programming in Chapter 11.

The linear programming problem considered in this section can be stated mathematically as:

Maximize

$$23X + 32Y + 18Z \qquad \text{linear objective function}$$

subject to the constraints

$$
\begin{aligned}
10X + 6Y + 2Z &\leqslant 2{,}500 \\
5X + 10Y + 5Z &\leqslant 2{,}000 \qquad \text{linear structural constraints} \\
1X + 2Y + 2Z &\leqslant 500
\end{aligned}
$$

(1.4)

and

$$X \geqslant 0, Y \geqslant 0, Z \geqslant 0 \qquad \text{nonnegativity constraints}$$

Our problem is to design an optimal program. What this means mathematically is that we must choose those values of X, Y, and Z that will maximize (1.1) *and* not violate the constraints given by (1.2) and (1.3). Note that the complete problem stated by (1.4) is nothing more than a mathematical statement of the situation confronted by the manager.

We now make some additional remarks regarding terminology.

1. The variables included in the linear objective function (e.g., X, Y, and Z), which we have termed activities or candidates, are also called the *decision variables*. The manager's task here is twofold: which variables to include in the objective function; and, what values to assign to these variables to optimize the objective function. The question of what the activities or the decision variables are is purely a matter of managerial judgment based on experience and market conditions. However, once the choice has been made as to what the decision variables in a given situation are, the problem of assigning values to these variables (i.e., how much to produce) is solved by the linear programming model.

2. The coefficients of the decision variables in the objective function are termed *profit* (for the maximization case) or *cost* (for the minimization case) *coefficients*. For example, the numbers 23, 32, and 18 are the three profit coefficients in (1.1). The information regarding these coefficients is obtained from the sales department.

3. The coefficients of the decision variables in the structural constraints are known as *input–output coefficients*. The prefix "input–output" is appropriate because the value of the coefficient here indicates the magnitude of

resource input required to achieve an output level of 1 unit. Thus, as seen in the first constraint of (1.2), 10 units of the cutting resource is needed to produce 1 unit of A. Similarly, the other input–output coefficients in (1.2) express the state of production technology of the problem. The information regarding input–output coefficients comes from such departments of the firm as the production department and research and development. The pertinent questions for the manager here relate to the reliability of information and possible future changes resulting from changes in technology.

4. The numbers on the right-hand side of the structural constraints express the maximum availability of resources ($\leqslant$), minimum* requirement of certain characteristics or variation thereof ($\geqslant$), or exact requirement ($=$). As seen from (1.2), all the constraints of our problem are of the "less than or equal to" type. The information regarding the magnitude of the capacity or requirement levels comes from a number of sources, such as the finance department, production department, and the quality-control department. Whether or not to acquire additional capacities of specified resources is an important managerial decision. As we shall see in Chapter 8, linear programming provides valuable assistance in making such decisions.

Finally, we observe, without proof, that if a linear programming problem can be solved, it can yield an *infinite* number of solutions.† This property will become evident when we present a graphical approach for analyzing and solving linear programming problems in Chapter 3. This property also means that if linear programming problems are to be solved in a finite number of steps, some *efficient* method of search must be available for identifying the optimal program. The simplex method, to be described in Chapter 7, is one of the efficient methods that have been developed for solving linear programming problems.

The remarks regarding some of the terminology employed in describing the linear programming problem, its assumptions, its three components, and the nature of linear programming solutions are only for introductory purposes. We shall elaborate on these as well as some other important aspects of linear programming in subsequent chapters.

1.3 EVOLUTION OF MODERN MANAGEMENT

In Section 1.1, we defined management as the process of integrating the efforts of a purposeful group or organization. This view emphasizes the decision-making aspects of management. That is, regardless of the type, nature, or level of maturity of the organization, the essence of management is to make decisions for the purpose of achieving individual, group, and organizational objectives. Managerial decisions relate to all aspects of organizational life, such as production, finance, marketing, research and development, cus-

---

* See, for example, the diet problem presented in Section 1.9.2.

† For an explanation, see Sections 3.3.3 and 4.3.

tomers, suppliers, stockholders, employees, motivation, morale, social responsibility, community relations, and government relations. These decisions have human, economic, social, political, psychological, and technological dimensions. Management, therefore, is at once a complex, dynamic, and social phenomenon which involves the sum total of all activities that relate to organizational life and organizational objectives. In this sense, the *practice* of management is as old as the history of man. Nevertheless, for purposes of providing a concrete starting point for describing the evolution of mangement, it is usually claimed that it was during the period of industrial revolution that the practice of management was first guided by *scientific* thought and produced any significant results. As a matter of fact, the very development of the factory system of production may be ascribed to the application of scientific management based on the principles of division of labor and specialization. In retrospect, we can also observe a parallel development in the *theory* of management, exemplified by the writings of Charles Babbage in the early part of the nineteenth century and followed by the work of men such as Frederick W. Taylor, Henri Fayol, Frank Gilbreth, and Henry Gantt.

Inevitably, with the passage of time, the developing theory of management and the scientific practice of management spurred the development of a systematic body of management knowledge. This gave birth to the *scientific management* movement. The interest generated by this movement led first to the development of the *scientific management* school, and eventually to other schools of management, such as *administrative management, human relations, behavioral science,* and *management science*. In a very real sense, as we show in Figure 1.1, modern management practice (and theory) is rooted in the thought processes that were set in motion by the scientific management school.

It is not our intention here to give a detailed history of the development of management thought. Instead, we shall list briefly the main elements of each school of thought (see Table 1.2) and then indicate that linear programming is the product of management science (or operations research) school.

The thrust of the scientific management school was to achieve specialization in terms of work, the worker, and the management by employing the scientific method* to study, analyze, and understand problems and phenomena. This school is associated with the contributions of Frederick W. Taylor (time study),† Frank Gilbreth (motion study), and Henry Gantt (control charts for scheduling and controlling production).

Taylor, Gilbreth, Gantt, and other pioneers of scientific management observed that, during most of the nineteenth century, considerable advances were made in the design and utilization of specialized machines and tools.

* The *scientific method* involves a systematic process of observation, problem definition, hypothesis formulation, testing, and arriving at a conclusion to guide action.

† Taylor is known as the founder of the scientific management movement. His famous work *The Principles of Scientific Management* was published in 1911. See Taylor [1911].

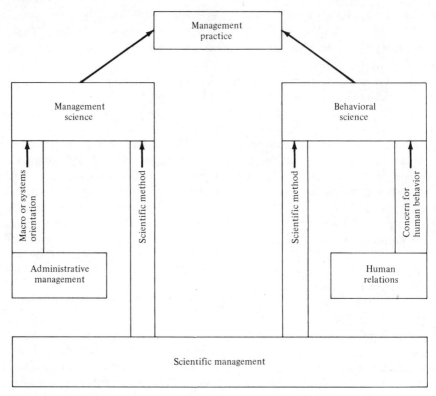

Figure 1.1   The Integrated Nature of Management Thought

This phenomenon, furthermore, had been accompanied by an increased productivity, on the one hand, and the development of larger business organizations, on the other. These pioneers soon recognized that, to gain further improvements in economic operations, the principle of specialization, already operating in the manufacturing activities, should be extended to the sphere of management activities. Taylor worked on this idea and developed what is known as the *functional organization.** This particular form of organization was based on the argument that if the worker was to get expert advice in the performance of his varied tasks, such advice, of necessity, must come from several different specialists, for there are natural limitations on the capabilities of any single person in terms of becoming a specialist in more than one area. Taylor, therefore, suggested that production activities be organized in terms of functions to be performed, and that each worker be under the supervision of, and accountable to, more than one foreman.

---

* In the specific organization plan advanced by Taylor, there were eight foremen exercising supervision over the workers. Of these, four foremen controlled the *planning activities* of time and cost, instruction, discipline, and order of work; the other four supervised the *performance activities* of speed, progress, inspection, and repair and setup.

Table 1.2. Schools of Management Thought

School of Management Thought	Major Contributors	Main Focus or Characteristics
Scientific management	Frederick Taylor Frank Gilbreth Henry Gantt	Scientific method applied to production problems Time study Motion study Functional organization
Administrative management	H. Fayol L. Urwick J. Mooney A. Riley	Management principles Macro orientation for administrative design Reliance on experience and intuition rather than empirical data
Human relations	Mary Parker Follett Elton Mayo	Importance of human-motivation "Group" approach to management Beginning of scientific experimentation on human problems
Behavioral science	Chris Argyris Rensis Likert Herbert Simon James March	Rigorous application of scientific method to individual, group, and organization behavior problems Emphasis on psychology, sociology, and anthropology for research in organization theory
Management science	P. M. S. Blackett George Dantzig C. West Churchman Russell Ackoff Richard Bellman	Mathematical models of management problems Scientific method Use of interdisciplinary teams Systems approach

Since this form of organization violated, in a sense, the principle of unity of command, it did not prove to be of lasting value in practice. However, it did provide the impetus for the development of the line-and-staff type of organization, which, although modified to accommodate special circumstances and technology, remains the predominant form of business organization today.

The purpose of the line-and-staff type of organization was, obviously, to provide the advantages of Taylor's functional emphasis without sacrificing the unity-of-command idea. Along with this new form of industrial organization, continuous efforts were made to evolve better methods for solving other management problems. As shown in Table 1.2, pioneers of the *administrative management* school made substantive contribution to the design and administration of organizations by developing a set of management principles* and by adopting a *macro* orientation to managerial problems, as opposed to the *micro* orientation of the scientific management school. Scientific management and administrative management provided the spark for several new developments in terms of improved tools, techniques, and methods for management practice. For example, an attempt was made to control inventories by employing the now well-known economic order quantity (EOQ) model. Improved procedures, based on scientific analysis and using mathematical and statistical concepts, were developed in such areas as personnel work, sampling inspection, quality control, work simplification, production, finance, and marketing. Newer and better methods of forecasting and budgeting were developed to better plan and control business operations. Considerable progress was also made in using mathematical and statistical techniques, such as calculus, design of experiments, analysis of variance, and Bayesian statistics to solve a host of business and industrial problems. These and similar developments continued to take place, during the 1920s, 1930s, and 1940s. However, neither the scientific management nor the administrative management school paid sufficient attention to psychological and sociological factors associated with human behavior. It was to fill this gap that first the human relations school and then the behavioral science school appeared on the scene.

The main thrust of the *human relations* school was to emphasize the importance of motivational factors to productivity and management. As a result, business managers began to introduce human relations programs in their companies based on the assumed, but not tested, premise that a happy worker was always a productive worker. The human relations school became popular in the 1940s and is considered the precursor of the behavioral science school. However, the human relations school has been criticized for two reasons. First, the proponents of this philosophy were perceived as "manipulative" persons who emphasized motivational factors only for superficial reasons. Second, it still lacked a genuinely scientific basis for inquiry into human behavior.

The introduction of new management techniques and the recognition by business managers of the importance of behavioral aspects of productivity resulted in the creation of *large* and *complex* business organizations. As a result there arose a new and complex class of managerial problems, "executive-type" problems. These problems, then as now, are the result of the need for reconciling the often-conflicting objectives of the different components of large organizations.

The increasing complexity of managerial decisions provided the impetus for the development of two specific aspects of management theory. The first

---

* Principles such as *unity of command* (an employee should receive orders only from one boss) and *scalar chain* (one should normally not short-circuit one's immediate superior) are two of the several general principles of management advanced by administrative management school.

related to the application of the scientific method for understanding human behavior in organizations. The concern here was to investigate, on a scientific basis, the psychological, sociological, and anthropological dimensions of management. As a result, several empirical studies were conducted to understand the behavior of individuals and groups within organizations under various kinds of economic, social, political, and technological environments. These studies form the central core of the *behavioral science* school.

The second aspect of management that needed careful study was the idea of a *systems*, as opposed to a functional, approach to management. The philosophy of the systems effort was that the goals and objectives of the separate parts of an organization are often in conflict; that the reconciliation of these conflicting goals and objectives is *the* executive function; and that such executive-type problems must be solved in terms of effectiveness of the *entire* system rather than its separate parts. The scientific research conducted with the philosophy of the systems approach has resulted in the evolution of *management science* (or operations research) school. Linear programming, as mentioned earlier, is a product of the management science school.

1.4 MANAGEMENT SCIENCE (OR OPERATIONS RESEARCH)

The availability of better management methods during the first four decades of this century brought profound changes to the business scene. Functional specialization, in both the technical and managerial fields, provided the means and opportunities for realizing further economies of scale in the production process. The net result was a phenomenal growth in the size of the business enterprise. Since business enterprises were organized into separate components (e.g., divisions, sectors, and departments), each with its own specific goals, there was, and under similar circumstances always will be, the possibility of potential conflict among the goals of the individual components. This state of affairs gave rise to *executive-type problems*, which consider the effectiveness of the organization as a whole in view of the conflicting goals of its component units. Let us illustrate by considering the problem of inventory control.

The question of inventory levels is a typical executive-type problem wherein different functional departments of a manufacturing enterprise have conflicting objectives. The sales department, in general, would like to have a large inventory of different products to achieve uninterrupted and satisfactory customer service. This clashes with the objective of the finance department, which would like to keep the inventory investment as low as possible. Similarly, the production department may wish to manufacture in large lot sizes to minimize production costs, but this would give rise to increased in-process inventories and higher working-capital requirements. The problem, therefore, is to develop an inventory policy that will minimize the total costs associated with the different and conflicting requirements of the functional departments. A number of similar examples can be given to emphasize the existence of multiple and often conflicting objectives for both organizations and individuals.

The terms *management science, operations research, systems sciences,* and *systems analysis* are often used interchangeably in the literature. Management science, as we know it today, made its start in the United Kingdom during World War II when a team of scientists was given the assignment of solving several complex military problems: selecting optimum gun sites; determining optimal convoy size, optimal depth for detonating antisubmarine charges, and optimal civil defense plans; and locating the most vulnerable spots in bombers. Management science pioneers used a research methodology that combined the inductive approach with the use of *analogy*. That is, wherever possible, an analogy with previously developed and tested logical structures was utilized in the process of model building.* The knowledge gained from wartime experiments was refined to arrive at a number of well-defined models (i.e., their general structure was precisely identified) dealing with problems of resource allocation, inventory control, queuing, routing, replacement, and so on. The attempt was to develop a variety of models, with known properties and deductively derived solutions, that can be matched and applied on a routine basis to certain types of recurring problems. This was indeed done, with increasing frequency and success in the 1950s and 1960s. Important contributors to promoting the philosophy and application of management science include C. West Churchman, Russell Ackoff, George Dantzig, Richard Bellman, and several others.†

What is management science and what do management scientists do? There is no unique answer to this question.‡ We consider management science as that branch of the field of management which employs a rational, logical, systematic, and scientific** approach in analyzing the process of management and management problems. To the extent possible, problems are examined with a systems orientation.§ And, in practice, management scientists develop scientific models that project the consequences of alternative courses of action and that incorporate the elements of chance, risk, and uncertainty to help managers make rational decisions and choose optimal policies.

Management science is both a body of knowledge and an approach for analyzing and solving management problems. As a *body of knowledge,*

* For a discussion of models, see Chapter 2.
† See Dantzig [1963, Chapter 2].
‡ See Loomba [1964, pp. 7–8].
** See Section 1.6.
§ See Section 1.5.

management science consists of various management theories, methods, models, and specific tools and techniques that can be used to handle a wide range of management problems. Management theories range all the way from individual, group, and organization behavior to theories of planning, control, and theories dealing with inventory, maintenance, and production scheduling. Management science models have been used to analyze wide areas of strategic, administrative, and operational problems. Linear programming, dynamic programming, stochastic programming, Markov chains, and simulation models provide some examples of management science models. Program evaluation and review technique, critical-path method, sensitivity analysis, different types of information models, and cost-effectiveness models are additional examples of topics that constitute the body of knowledge in management science.*

As an *approach*, management science refers to the attitude with which management scientists view, analyze, and solve management problems. The approach is that of a man of science grounded in the discipline of inductive as well as deductive inference, model building, and theory construction. The essence of this approach is, first, that problems must be expressed quantitatively and, second, that symbolic (as opposed to verbal) modes of expression and reasoning are to be preferred.

The focus of management scientists has mainly been on the economic–technical subsystems of organizations. Management science is essentially normative in nature—that is, it *prescribes* a specific managerial course of action. It tells the manager how he *should* behave. It outlines specific solutions to specific problems based on a set of assumptions. For example, if the problem presented in Section 1.2 is solved by the simplex method of linear programming, the *optimal* solution will be identified. The implication is that the manager *should* proceed to implement the optimal solution.†

Management science has been effective in two ways. First, the scientific approach has yielded dividends in improving the *art* of management. Second, several breakthroughs have been made in isolating and solving complex decision problems at the operational level. We refer here to such problems as scheduling, inventory, facilities location, distribution, queueing, replacement, maintenance, design, and information systems.

Management science and the advent of high-speed computers are perhaps two of the most important developments of the period following World War II. It may be mentioned that the parallel development of management science and computers is not a coincidence. One could have only a limited value without the other. It is the availability of high-speed computers that has made the economic solution of large-scale problems a reality. Computers have facilitated the analysis and prediction of the future behavior

* See Levey and Loomba [1973, Chapters 12 and 13] for a brief description of the various management science models.

† We emphasize the normative aspects because, in practice, the manager might, because of political considerations, deviate from the solution suggested by the model.

of complex production, marketing, financial, and other systems by making the use of analytical, as well as simulation models, economically viable.*

We conclude this section by listing the four main characteristics of management science: (1) systems approach, (2) application of the scientific method, (3) interdisciplinary teamwork, and (4) the use of models.

1.5 THE SYSTEMS APPROACH

The *systems approach* means that the manager makes a *conscious* attempt to understand the relationships between various parts of the organization and their role in supporting the overall performance of the organization. Before solving a problem in any functional area, or at any organizational level, or in any specific sector of the organization, the manager must understand fully how the overall system will respond to changes in its component parts. In short, the systems approach is based on the conviction that before implementing any functional solution, one must examine its ultimate effect on the system. Furthermore, the process of problem formulation and definition at lower levels of the organization must, if at all possible, fit into the boundaries defined by higher-level objectives. This implies comprehensiveness, both in terms of objectives and problem formulation.† Thus, in linear programming problems, once the system is defined, the objective function relates to the entire system and not the separate parts of the system. The conflict among competing activities is, in fact, resolved through the device of the overall objective function.

1.6 APPLICATION OF THE SCIENTIFIC METHOD

The scientific method involves a systematic process of observation, problem definition, hypothesis formulation, testing, and arrival at a conclusion to guide action. If the problem to be solved is of the *repetitive* type, the scientific method can yield a generalized framework, or a model, that can be used to solve similar problems in different settings. This process of developing

---

* Analytical models have specified mathematical structure, and they can be solved by known analytical or mathematical techniques. For example, linear programming is an analytical model. Simulation models also have a mathematical structure, but they cannot be solved by mathematical techniques. See Section 2.3.

† However, practical consideration of time, information, cost, and feasibility often force the manager to solve parts of the problem individually and in sequence. This is in contradiction to the idea of overall optimization, which is the goal of management science. But suboptimization (suboptimization occurs when there are conflicting goals and objectives, or imperfect information) is a fact of life, and the importance of management science lies in understanding when it is necessary and when it can be avoided.

generalized models is known as the *inductive* approach to model building. It is also possible to develop general models through the use of a *deductive* approach. That is, based on certain assumptions (e.g., linearity, certainty, continuous variables), the general models are developed by applying the rules of mathematics. Both approaches to model building have been used in developing management science models. Once the general model is developed, it can be used to predict the effect of changes on the system behavior. For example, linear programming is a general model that can be applied to solve certain types of problems. Also, it can be used to predict changes in system behavior (e.g., profit) as a result of changing profit coefficients, input–output coefficients, resource capacities, and so on.

1.7 INTERDISCIPLINARY TEAMWORK

The idea of *interdisciplinary teamwork* is necessary in solving complex management problems because of the advanced state of specialization in many disciplines. The body of specialized knowledge has grown to a point where it is impossible for one person to be a specialist in more than one branch of a scientific discipline. And the problems of modern-day life are multidimensional, not single-dimensional. Furthermore, the same problems can be viewed, with beneficial results, by men from different disciplines. It has been claimed that "there are no such things as physical problems, biological problems, social problems, psychological problems, economic problems, and so on. There are only problems; the disciplines of science represent different ways of looking at them."*

Large and complex problems require that they be subjected to analysis by a team of specialists who represent a wide range of skills. This multi-disciplinary view produces cross-fertilization of ideas, takes advantage of accumulated knowledge, and makes certain that all ramifications of the problem have been considered. Many organizational problems, for example, have economic, social, political, engineering, physical, biological, and psychological aspects. Although it is impossible for one person to specialize in such a range of disciplines, it is conceivable that a team can be organized so that each particular aspect of the problem could be analyzed by a specialist in that field. By pooling their specialized talents, the team can develop better and advanced solutions to old problems and new solutions to new and complex problems. The scientific mind from each discipline attempts to abstract the essence of the problem and relate it structurally to similar problems from his own field. If there is a structural similarity between the new problem and the one familiar to the scientist, there is the possibility of applying the old and tested solution methods for solving new problems. When

* Ackoff and Sasieni [1968, p. 7].

the entire team is attempting to develop analogies in this manner, the possibility of finding a solution increases markedly.

The idea of interdisciplinary teamwork is important because in the formulation and solution of managerial problems, inputs are needed from diverse sources. Linear programming problems are no exception.

1.8 THE USE OF MODELS

A *model* is a particular representation of reality. We emphasize the "particularity" in models because managers can perceive the same problem differently, depending upon their particular interests and attention focus. The management science or quantitative models, as opposed to verbal models, have the advantage that, given the same assumptions, different persons will arrive at the same model. And, regardless of the method of solution, a problem with well-specified structure and assumptions will have the same optimal solution. For example, whether we use the graphical, algebraic, or the simplex method, the optimal solution to the problem described in Section 1.2 is the same.*

A model can be represented by different means: verbal, physical, graphical, mathematical, and so on. The use of mathematical or quantitative models is the core of management science. Linear programming is a mathematical model.

The concepts of models, model building, and model implementation are very important in managerial decision making. We discuss these concepts in Chapter 2.

1.9 SELECTED APPLICATIONS OF LINEAR PROGRAMMING

As mentioned earlier, the linear programming model can be applied to *any* problem in which the objective is to optimize a linear objective function, subject to a set of linear and nonnegativity constraints. The model is, therefore, quite general and has successfully been applied to a host of real-life problems. In this section we present a few examples that are considered classical in the practical application of linear programming.

---

* The graphical, algebraic, and simplex methods are discussed in Chapters 3, 4, and 7, respectively.

1.9.1   The Product-Mix Problem

An example of the product-mix problem was presented in Section 1.2. A product-mix problem exists whenever the manager is to determine an optimal production program or product mix (which products to produce and in what quantities). Such a problem can be solved by linear programming if the objective is to maximize a linear objective function, subject to a set of linear and nonnegativity constraints. Typically, in a product-mix problem, the inequality constraints are of the "less than or equal to" type.*

The product-mix problem is a problem that arises in all types of business situations. Portfolio-selection and media-selection problems (Sections 1.9.5 and 1.9.6) are, in effect, examples of the product-mix problems in the areas of finance and marketing.

1.9.2   The Diet Problem

The diet (or feed-mix) problem arises whenever the manager is to determine that optimal combination (or program) of diet, or feed mix, which will satisfy specified nutritional requirements and minimize the total cost of purchasing the diet. Data for a typical diet problem are given in Table 1.3.

Table 1.3

Nutrient	Food A	Food B	Weekly Nutritional Requirement
I	2	3	3,500
II	6	2	7,000
Cost per unit of food (cents)	10.0	4.5	

In the diet problem, the objective is to minimize a linear objective function, subject to a set of linear and nonnegativity constraints. Typically, in the diet problem, the constraints are of the "greater than or equal to" type, because some minimum quantities of specified nutrients must be supplied by the

---

* This is because we have to *maximize* a linear objective function. What, for example, would be the practical consequence of a linear programming problem in which the linear function is to be maximized and *all* constraints are of the "greater than or equal to" type?

diet.* The problem given in Table 1.3 can be mathematically stated as follows:

Minimize $\qquad 10X + 4.5Y$

subject to the constraints

$$\left.\begin{array}{l} 2X + 3Y \geqslant 3,500 \\ 6X + 2Y \geqslant 7,000 \end{array}\right\} \tag{1.5}$$

and

$$\left. X, \quad Y \geqslant \quad 0 \right\}$$

1.9.3 The Transportation Problem

The specifications of a typical transportation problem are given in Table 1.4. In the transportation problem, we are required to design a program of transportation (how much to ship from a set of origins, or sources, to a set of destinations), of a homogeneous commodity, so as to minimize the

Table 1.4

Origin	Destination			Monthly Supply
	D_1	D_2	D_3	
O_1	8 x_{11}	5 x_{12}	7 x_{13}	300
O_2	4 x_{21}	7 x_{22}	6 x_{23}	250
Monthly Demand	150	175	225	550

total costs of transportation. Here, typically, the constraints specify the exact level of supply available at the origins and the exact level of demand that is required by the destinations. Also known is the cost of shipping 1 unit from each origin to each destination. These costs are assumed to stay constant during the planning horizon and are usually entered in the upper right-hand corners of the cells, as shown in Table 1.4.

---

* The "greater than or equal to" type inequalities are a natural consequence of the situation in which a linear function is to be minimized. What, for example, would be the practical consequence of a linear programming problem in which the linear function is to be minimized and *all* constraints are of the "less than or equal to" type?

The decision variables in the transportation problem are identified by double subscripts. For example, the quantity to be shipped from the first origin to the second destination will be denoted by x_{12}. Mathematically, the transportation problem given in Table 1.4 can be stated as follows:

Minimize $8x_{11} + 5x_{12} + 7x_{13} + 4x_{21} + 7x_{22} + 6x_{23}$

subject to the constraints

$$\left. \begin{array}{l} x_{11} + x_{12} + x_{13} \qquad\qquad\qquad\qquad = 300 \\ \qquad\qquad\qquad x_{21} + x_{22} + x_{23} = 250 \\ x_{11} \qquad\qquad\quad + x_{21} \qquad\qquad\qquad = 150 \\ \qquad x_{12} \qquad\qquad\qquad + x_{22} \qquad\quad = 175 \\ \qquad\qquad x_{13} \qquad\qquad\qquad\quad + x_{23} = 225 \end{array} \right\} \quad (1.6)$$

and

$$x_{11}, \quad x_{12}, \quad x_{13}, \quad x_{21}, \quad x_{22}, \quad x_{23} \geqslant 0$$

The transportation problem is discussed in Chapter 9. The transportation problem has a special structure, as can be ascertained from (1.6). Note that all the input–output coefficients in a transportation problem are either 1 or 0.

1.9.4   The Assignment Problem

The specifications of a typical assignment problem are given in Table 1.5, where the cost (or profit) of assigning each machine to each job is shown. The objective in the assignment problem is to determine an assignment program (which machine should be assigned to which job) so as to minimize total costs or maximize total profits. Note that the assignment problem involves the matching of *exactly one* machine to *exactly one* job. In this sense, it is a special case of the transportation problem, in which the supply available at a specific origin could be shared by several destinations. In the assignment model the problem is so formulated that the available supply at each origin, and the required demand at each destination, is exactly 1.

Table 1.5. Cost of Assignments

Job	Machine		
	M_1	M_2	M_3
J_1	4	2	6
J_2	5	7	8
J_3	9	7	6

Thus, while making the assignments of origins to destinations, the one-to-one match is either made or it is not made. This means that each decision variable must take a value of either 1 or 0. We shall discuss the assignment model in Chapter 10.

1.9.5   The Portfolio-Selection Problem

A portfolio-selection problem arises whenever a given amount of money is to be allocated among several investment opportunities. The portfolio-selection problem is quite general in the sense that it can exist for individuals, trust-fund managers, or corporations. The components of the portfolio can be stocks, bonds, savings certificates, projects, companies to be acquired, or plants to be opened. The portfolio-selection problem, like other allocation problems, can be formulated at different levels of mathematical sophistication and solved by applying different methods of solution. For example, the objective function to be maximized can be nonlinear rather than linear.* A single-period allocation problem can be solved by dynamic programming.† In this section we present a simplified portfolio-selection problem.

Assume that the management of Atlas Corporation wants to invest, for a specified period, a sum of $1,000,000. The finance department of the company is asked to provide data regarding expected yields from good-quality common and preferred stocks, corporate bonds, government bonds, and saving certificates. The relevant information regarding expected annual yields of different categories of investment is shown in Table 1.6.

Table 1.6

Category of Investment	Expected Annual Yield (per cent)
Common stocks	5
Preferred stocks	7
Corporate bonds	10
Government bonds	8
Saving certificates	6

Assume that these returns are stable and that, for the planning horizon, they will remain constant.‡ Assume further that the diversification goals of the management are specified by these statements (or constraints):

---

* See Markowitz [1952].

† For a simple illustration, see Loomba and Turban [1974, p. 369].

‡ This is a bold assumption. In practice, the manager must attach some sort of risk factor to each category of investment. A linear programming formulation that takes this modification into account is easily obtainable. See Problem 1.12 at the end of the chapter.

1. Investment in common and preferred stocks should not be more than 30 per cent of the total investment.
2. Investment in government bonds should not be less than the investment in saving certificates.
3. The investment in corporate and government bonds should not be more than 50 per cent of the total investment.

Our problem, then, is to select the investment portfolio that will maximize expected annual return, and simultaneously meet the diversification and other constraints.

In formulating this problem, we introduce a slight variation in defining the activities. That is, here we define activities, or the decision variables, as percentages, rather than as quantity levels. Thus we define x_1, x_2, x_3, x_4, and x_5 as percentages of the total portfolio (or investment) to be allocated, respectively, to common stocks, preferred stocks, corporate bonds, government bonds, and saving certificates. The problem can now be stated mathematically as follows:

Maximize $\quad 0.05x_1 + 0.07x_2 + 0.10x_3 + 0.08x_4 + 0.06x_5$

subject to the constraints

$$\left.\begin{aligned}
x_1 + x_2 &\leqslant 0.30 \\
x_4 - x_5 &\geqslant 0.0^* \\
x_3 + x_4 &\leqslant 0.50 \\
x_1 + x_2 + x_3 + x_4 + x_5 &= 1.00
\end{aligned}\right\} \quad (1.7)$$

and

$$x_1, \quad x_2, \quad x_3, \quad x_4, \quad x_5 \geqslant 0$$

The equality constraint in (1.7) represents the fact that our total investment must equal 100 per cent.

1.9.6 The Media-Selection Problem

The DISCO Manufacturing Company has decided to spend $200,000, during the next fiscal year, to advertise its electrical ovens. The management of DISCO is in the process of designing an advertising program that will use one or more of these four media: television, radio, newspapers, and magazines. Their marketing research department recommends that the allocation of advertising budget be made in such a way that the exposure, E, of the company's product is maximized. Past experience and records indicate the data in terms of exposure per dollar spent in advertising (Table 1.7). Also, as shown in the table, the management has specified certain maximum expenditures in each media. Assume further that the combined

---

* Statement 2 of the problem can be expressed as $x_4 \geqslant x_5$ or $x_4 - x_5 \geqslant 0.0$.

Table 1.7

Media	Exposure per Dollar Spent in Advertising	Maximum Expenditures
Television	7	100,000
Radio	4	20,000
Newspapers	5	50,000
Magazines	3	30,000

expenditures for newspapers and magazines should not be less than the allocation for radio. Let x_1, x_2, x_3, and x_4 represent the dollar amounts to be expended on television, radio, newspapers, and magazines, respectively. Then our media-selection problem can be mathematically stated as follows:

Maximize $\quad E = 7x_1 + 4x_2 + 5x_3 + 3x_4$

subject to the constraints

$$
\begin{aligned}
x_1 &\leqslant 100{,}000 \\
x_2 &\leqslant 20{,}000 \\
x_3 &\leqslant 50{,}000 \\
x_4 &\leqslant 30{,}000 \\
-x_2 + x_3 + x_4 &\geqslant 0^*
\end{aligned}
\tag{1.8}
$$

and

$$x_1, \quad x_2, \quad x_3, \quad x_4 \geqslant 0$$

1.9.7 The Blending Problem

The blending problem is very much like the diet problem. It is, however, not as completely stated. Typically, the idea is to produce a blend out of specified commodities or constituents whose characteristics and costs are given. Minimum acceptable characteristics are given and are specified in terms of percentages rather than in absolute amounts. The objective function to be minimized is formulated not in total cost, but in terms of such measures as cost per gallon or cost per pound of alloy. In an alloy-blending problem, we may be required to have stated percentages of certain metals, such as copper, tin, or zinc, to produce an alloy. The information regarding the metal contents and unit costs of certain alloys available in the market is given. The problem, then, is to design a blend that minimizes cost per pound and also satisfies the stated requirements. Table 1.8 specifies such a problem.

---

* The requirement that the combined expenditures for newspapers and magazines should not be less than the allocation for radio can be expressed as $x_3 + x_4 \geqslant x_2$ or $-x_2 + x_3 + x_4 \geqslant 0$.

Table 1.8

Metal Content	Alloy Available in the Market			Required Blend
	A	B	C	
Copper	0.3	0.1	0.7	0.20
Tin	0.5	0.6	0.1	0.53
Zinc	0.2	0.3	0.2	0.27
Cost per pound	$13	$11	$9	

Let x_1, x_2, and x_3 represent, respectively, the percentages of alloys A, B, and C in 1 pound of the blended alloy. It is then clear that one of the constraints must be $x_1 + x_2 + x_3 = 1$. The remaining information is quite clear from Table 1.8. Mathematically, the problem can be stated as follows:

Minimize $\qquad 13x_1 + 11x_2 + 9x_3$

subject to the constraints

$$
\left.
\begin{aligned}
0.3x_1 + 0.1x_2 + 0.7x_3 &= 0.20 \\
0.5x_1 + 0.6x_2 + 0.1x_3 &= 0.53 \\
0.2x_1 + 0.3x_2 + 0.2x_3 &= 0.27 \\
x_1 + x_2 + x_3 &= 1
\end{aligned}
\right\} \quad (1.9)
$$

and

$$x_1, \quad x_2, \quad x_3 \geqslant 0$$

A second variation of the blending problem results if we are to design an optimal mix of production runs that consume *inputs per run* (as opposed to input per unit of output) and produce specified outputs of *joint* products per run.[*] Consider, for example, the problem of designing an optimal mix (in terms of runs) of two production processes in an oil refinery. The refinery uses two production processes (I and II) to produce two types of gasoline (Regular and Premium). Each run of process I requires 3 units of crude A and 8 units of crude B to produce 6 units of Regular gasoline and 4 units of Premium gasoline. Each run of process II requires 6 units of crude A and 4 units of crude B to produce 5 units of Regular gasoline and 7 units of Premium gasoline. The availability of supply of crudes A and B is 300

---

[*] Several other variations of the blending problems exist. For another variation see Problem 7.17. Yet another variation is presented in Problem 7.18, which is a prototype of a simple production planning problem.

Table 1.9

Production Process	Input per Production Run (units)		Output per Production Run (units)	
	Crude A	Crude B	Regular Gasoline	Premium Gasoline
I	3	8	6	4
II	6	4	5	7
	300	400	200	150
	Input supply		Output requirement	

and 400 units, respectively. The refinery must produce 200 units of Regular gasoline and 150 units of Premium gasoline. The problem data are organized in Table 1.9. Assume that the profits per production run from process I and II are \$10 and \$15, respectively. Let x_1 and x_2 be the number of production runs of process I and II, respectively. Our problem can then be mathe-matically stated as follows:

Maximize $\qquad 10x_1 + 15x_2$

subject to the constraints

$$\left.\begin{array}{l} 3x_1 + 6x_2 \leqslant 300 \\ 8x_1 + 4x_2 \leqslant 400 \end{array}\right\} \text{ supply constraints}$$

$$\left.\begin{array}{l} 6x_1 + 5x_2 \geqslant 200 \\ 4x_1 + 7x_2 \geqslant 150 \end{array}\right\} \text{ output constraints}$$

$$x_1, \qquad x_2 \geqslant 0$$

1.10 SUMMARY

In this chapter we have introduced the reader to the nature and scope of linear programming. To provide a proper managerial perspective, we described briefly the evolution of modern management. The main focus and the major characteristics of various schools of management thought were briefly described. Since linear programming is the product of the management science (operations research) school, we discussed the philosophy and characteristics of management science. In particular, the ideas of the systems approach, the scientific method, interdisciplinary teamwork, and the use of models were explained. The meaning of such terms as *activity*, *program*, *input–output coefficients*, and *optimal program* was illustrated with reference to a specific problem. We also stated a set of assumptions that underlie linear programming: certainty, linearity, continuous variables, and static conditions.

Linear programming is but one of the many decision models that have been developed in the area of management science. How does linear programming differ from some of the other decision models? What are the essentials of a basic decision model, and how do they relate to the linear programming model? We provide answers to these questions in Chapter 2.

REFERENCES

Ackoff, R. L., and M. W. Sasieni. *Fundamentals of Operations Research*. New York: John Wiley & Sons, Inc., 1968.

Dantzig, G. B. *Linear Programming and Extensions*. Princeton, N.J.: Princeton University Press, 1963.

Levey, S., and N. P. Loomba. *Health Care Administration*. Philadelphia: J. B. Lippincott Company, 1973.

Loomba, N. P. *Linear Programming*. New York: McGraw-Hill Book Company, 1964.

Loomba, N. P., and E. Turban. *Applied Programming for Management*. New York: Holt, Rinehart and Winston, Inc., 1974.

Markowitz, H. "Portfolio Selection." *Journal of Finance*, **7**, No. 1 (Mar. 1952), 77–91.

Taylor, F. W. *Principles of Scientific Management*. New York: Harper & Row, Publishers, 1911.

REVIEW QUESTIONS AND PROBLEMS

1.1. Define the term *programming*. What are the various assumptions of a linear programming model? Discuss them by using the example of a product-mix problem.

1.2. What are the various fields in which linear programming has been successfully used in making managerial decisions? Give an example of each.

1.3. Explain the term *linear* and its implications with reference to a linear programming model.

1.4. Briefly describe the evolution of management thought.

1.5. What areas of management were given less attention by the proponents of scientific management and administrative management?

1.6. In what way does the method of attacking a problem in management science differ from that followed in the administrative management school?

1.7. What are the four most important characteristics of management science? Explain in your own words the significance of the scientific method, interdisciplinary teamwork, and the use of models.

1.8. Describe the systems approach. Explain why the use of the systems approach is paramount in a modern corporation.

1.9. What was Frederick Taylor's idea of "functional organization"? What problems were inherent in his system?

1.10. Define a model. List some important management science models. What are some of the special characteristics of management science models?

1.11. What is the basic structure of a linear programming problem? What factors should be considered before a problem can be put into the linear programming format?

1.12. Assume that the management of Scarsdale Mutual, Inc., wants to invest a sum of $1,000,000. The finance department has provided the following data regarding the yields and the risk factor (beta factor) for the securities:

Category of Investment	Expected Annual Yield	Risk or Beta Factor
Common stocks type A (utilities)	5	1
Common stocks type B (blue chip)	8	1.6
Corporate bonds	10	0.5
Government bonds	8	0
Saving certificates	6	0

Assume that the yields and risk factors remain constant for the planning horizon. Cast the problem in the linear programming format. The constraints are as follows:

(a) The investment in government bonds and saving certificates should not be more than 50 per cent of the total investment.

(b) Common stocks should not constitute more than 40 per cent of the portfolio.

(c) The beta factor of the portfolio should be less than or equal to 1.

1.13. A company currently uses two processes to produce three grades of plywood. Each time process I is run for 1 hour, the output consists of 4 units of Grade A, 5 units of Grade B, and 6 units of Grade C. Each time process II is run for 1 hour, the output consists of 6 units of Grade A, 5 units of Grade B, and 2 units of Grade C. There are available 400 hours of Process I and 600 hours of Process II. The company wants at least 4000 units of Grade A, 4000 units of Grade B, and 3000 units of Grade C to be produced. The objective is to find the minimum cost to the firm if it costs $10 per hour for process I to be run and $12 per hour for process II to be run. Formulate this problem as a linear programming model.

2
Models and Linear Programming

2.1 INTRODUCTION

Linear programming is a *decision model*. It is important, therefore, that we understand the meaning and purpose of models and the nature and role of decisions and decision making. Both of these topics, models and decision making, are extremely broad, complex, and multidisciplinary topics. Our purpose here is to present the basic ideas of models and decision making and relate these ideas to linear programming. This will enable us to understand the place, potential, and limitations of linear programming in the broad field of management science (operations research).

In this chapter we first discuss the idea of a model, the purpose of a model, and the concept of model building. We then present some classifications of models, based on various criteria of classification. This is followed by a brief description of some well-known models of management science. Finally, we examine some aspects of decision making, present a basic decision model, and relate the essentials of the basic decision model to linear programming.

2.2 MODELS AND MODEL BUILDING*

The logical analysis of all decision problems is based on the concepts of models, model building, and model implementation. A model is a *particular* representation of reality. For example, a balance sheet that gives, at a given point in time, an instant picture of assets, liabilities, and net worth of a corporation is a financial model. Similarly, an income statement that lists, for a given period of time such as 1 year, revenues, expenditures, and net income of a corporation is also a financial model. A balance sheet and an annual income statement are thus examples of models that represent some aspects of the financial health of a corporation. Similarly, we can build quantitative or mathematical models of several types of managerial problems related to production, marketing, or other areas.

We emphasize the "particularity" of models because managers can perceive the same problem differently, depending upon their particular interests and attention focus. The management science or quantitative models, as opposed to verbal models, have the advantage that, given the same assumptions, different persons will arrive at the same models. Furthermore, application of these models to problems of specified properties will yield the same solutions.

A model embodies our attempt to represent reality, and it is built with the purpose of enhancing our understanding of reality. The reality, however, is so complex that often we cannot visualize and understand it completely. Furthermore, even if we do understand it, our attempts at representation may succeed only partially, because the tools and means needed to adequately represent reality may not be fully developed or available, and also because of the existence of uncertainties. Thus, by definition, there are some inherent weaknesses in the use of models to represent reality. This does not mean, however, that we should reject the approach of model building. On the contrary, analytical model building is not only sound but the best available approach to the study of decision problems. It is much easier, less costly, and less time consuming to obtain information from models than from experimentation with the reality that the model represents. Furthermore, the main advantage of a tested and proved model is that once a general model is set up, it can be used to solve a whole class of similar problems. All the manager has to do is recognize the nature of the given problem and see if it can usefully be represented by a particular proved model. For example, we realize that all maximization and minimization problems that have the characteristics of a strictly defined linear programming problem can be solved by the general model of the simplex method.

The purpose of a model is to *describe, explain, predict,* or *prescribe* the performance of a system. To be most useful, the model should explain and

* Sections 2.2 to 2.6 are adapted from Levey and Loomba [1973, Chapter 4].

predict the behavior of individual components of a system, the response of the system to changes in one or more of its components, and the effect on the system of internal as well as external disturbances. After the model has been constructed, it should be tested for applicability. However, a model need not be 100 per cent applicable. The real issue is the degree of usefulness of the model. Thus the assumption of linearity in linear programming models may not be completely faithful to the realities of business problems, but they are nevertheless extremely useful in solving certain types of business and managerial problems.

To build a model, we have to use some device to represent an object or subject of inquiry. This device may be physical, schematic, symbolic, or mathematical, or a combination of these. Linear programming, for example, is a mathematical model whose three components were identified in Chapter 1. In building a linear programming model, we must systematically identify the decision variables, profit or cost coefficients, input–output coefficients, resource capacities, and other requirements. Then, having satisfied ourselves regarding such assumptions as linearity, certainty, continuous variables, and static conditions, we proceed to build a linear objective function, linear structural constraints, and other constraints and relationships. Once the model is built, it can be solved by a known method that is effective. For example, as we shall describe later, transportation problems can best be solved by utilizing not the general simplex method (Chapter 7) but the MODI technique (Chapter 9). The cost effectiveness of models is of paramount concern to managers.

2.3 CLASSIFICATION OF MODELS

We can establish several different classifications of models, depending upon the purpose or criterion of classification. Figure 2.1 presents a classification of models according to the *degree of abstraction*. Any three-dimensional model that looks like the real thing but is either reduced in

Figure 2.1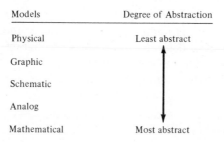

size (e.g., toy airplane) or scaled up (e.g., plastic model of the human heart) is a *physical model*. Physical models are easy to observe, build, and describe, but they are difficult to manipulate and not very useful for prediction. However, they are quite helpful in the area of physical planning. For example, three-dimensional models of plant layouts, housing developments, and city planning are often employed for improving the detail and quality of plans.

An organization chart is a *graphic* (block-type) *model* that depicts the intended system of organizational authority–responsibility relationships. The flow process chart showing what happens (operation, storage, delay, inspection, etc.) at different stages during the complete processing of a product is a *schematic model*. The main features of a computer program are often represented by a schematic description of steps that connect the start to the end of the computer program. Graphic and schematic models are extremely useful in providing a visual picture of the system under study. A picture, as they say, is worth a thousand words.

Analog models represent a system (or object of inquiry) by utilizing a set of properties different from that which the original system possesses. For example, an analog computer is the physical (mechanical or electrical) representation of the variables in a problem. Various colors on a map may represent water, desert, continents, and so on; or they may represent military alliances among nations. Similarly, a map is an analog model which shows roads, highways, towns, and their interrelationships. Graphic, schematic, and analog* models are easier to manipulate and more general than physical models.

Mathematical, or *symbolic models* represent systems (or reality) by employing mathematical symbols and relationships. Mathematical models are precise, most abstract, general rather than restricted, and can be manipulated easily by utilizing the laws of mathematics. The mathematical model for any straight line, for example, is $y = a + bx$, where a and b are, respectively, the intercept and slope of the line. Any specific straight line can be represented by assigning numerical values to the parameters a and b.

A second classification of models, suggested by Forrester [1961], is shown in Figure 2.2. Forrester's classification is useful in understanding the nature and role of models to represent management and economic behavior of organizations. Also, we have shown by a contour in Figure 2.2 the domain of linear programming covered in this book.

Models can also be classified as *deterministic* or *probabilistic models*. Deterministic models assume conditions of complete certainty and perfect knowledge. Linear programming, transportation, and the assignment models are examples of deterministic models. Probabilistic models handle those situations in which the consequences or payoffs of managerial actions cannot be predicted with certainty. For example, if we toss a coin, we cannot

---

* The graphic, schematic, and analog models are of the same species—the difference is one of degree rather than kind.

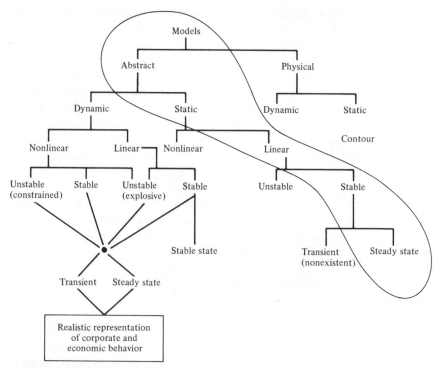

Figure 2.2   Classification of Models
(From J. W. Forrester, *Industrial Dynamics*, Cambridge, Mass.: The MIT Press, 1961, p. 49.)

predict with certainty whether the outcome will be a head or a tail. However, it is possible to forecast a pattern of events, based on which managerial decisions can be made. For example, insurance companies are willing to insure against risks of fire, accidents, sickness, and so on, because the pattern of events has been compiled in the form of probability distributions. We do not treat probabilistic models in this text.

Models can also be classified by *purpose*. For example, the purpose of a model may be to describe, explain, predict, or prescribe. In the process of model building and theory construction, we usually progress from the initial descriptive stage to the final prescriptive stage. The explanatory and the predictive stages are the intermediate stages. A descriptive model simply describes what *is*, based on observation, survey, questionnaire results, or other available data. The results of an opinion poll represent a descriptive model. Attempts are then made to explain the model and make predictions regarding certain events. For example, based on survey results, or early returns of an election, television networks attempt to explain and predict the election outcome *before* all the votes are actually counted. Finally, when a predictive model has been repeatedly successful, it can be used to *prescribe* a course of action. Linear programming is a prescriptive (also called normative) model, because it prescribes what the manager *ought* to do.

Models can also be classified according to their *form* and *content*. The form refers to the structure of a model, the relationship among its various components, and its general characteristics. For example, the form of linear programming problem is specified by its three components and its various assumptions. The content of a model refers to the specific context or situation to which the model is being applied. For example, a linear programming model can be used to solve specific problems in production, finance, marketing, and so on. In Section 2.4 we present a classification of models based on form.

Models can also be classified as *static* or *dynamic*. Essentially, static models do not consider the impact of changes that take place during the planning horizon. That is, they are independent of time. Also, in static models, only one decision is needed for the duration of a given time period. It should also be noted here that linear programming is a static decision model. Dynamic models consider time as one of the important variables and admit the impact of changes generated by time. Also, in dynamic models, not one, but a series of decisions, is required during the planning horizon. We do not treat dynamic models in this book.*

An important classification of models is that of *analytical* versus *simulation models*. Briefly, analytical models have specified mathematical structure and they can be solved by known analytical or mathematical techniques. For example, the general linear programming model, as well as the special-structure transportation and assignment models (see Chapters 9 and 10), are analytical models. Simulation models also have mathematical structures, but they cannot be solved by applying purely the tools and techniques of mathematics. A simulation model is essentially computer-assisted experimentation on a mathematical structure of a real-life system in order to study the system under a variety of assumptions. Probabilistic systems are often best analyzed through simulation experiments.†

2.4 MANAGEMENT SCIENCE (OPERATIONS RESEARCH) MODELS

We present in this section six classes of management science models. Each class is characterized by a specific structure or form, and each represents a category to which management scientists have made the most important contributions during the last few decades. The following six classes of models

---

* For a brief discussion of dynamic models, see Loomba and Turban [1974, Chapter 8].
† For a discussion of simulation models, see Levey and Loomba [1973, Chapter 12].

will be briefly described:

1. Allocation models.
2. Inventory models.
3. Queueing models.
4. Replacement models.
5. Maintenance models.
6. Competitive models.

2.4.1 Allocation Models

Allocation models are used to solve that class of problems in which a number of candidates or activities are competing for limited resources. The resources are limited in two ways. First, in a given time period, there may be a limit on the quantity, beyond which resources cannot be purchased or employed. Second, they may be limited in the sense that within the boundaries of the problem each candidate usually cannot be allocated the given resources in the most efficient way on an individual basis. For example, let us assume that during a particular time period a firm is planning to manufacture three different products, X, Y, and Z, each yielding a specific contribution to overhead and profit. Assume, further, that two different machine processes (resources) are required in manufacturing each of these products. The allocation problem is to determine a program of production, or a product mix, that will maximize not the *individual* contribution of a given product but the *overall* effectiveness of the production program. Mathematical programming, linear and nonlinear, is one of the methods used in solving allocation problems. The scope of mathematical programming can be appreciated by a classification developed by Dantzig [1963], as shown in Figure 2.3.

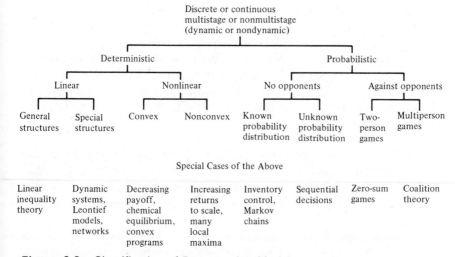

Figure 2.3 Classification of Programming Models

2.4.2 Inventory Models

Inventory models deal with that general class of problems in which something is stored to meet future demand. Essentially, inventory models answer the questions of "How much?" and "When?" regarding the quantity and time of procurement and/or production of inventory items. Associated with any inventory problem are certain categories of costs, such as ordering costs, holding costs, out-of-stock costs, and costs of inventory items. As the level of average inventory over a given time period increases, some of these costs increase while others decrease. For example, for a known demand rate and a known lead time, and with no discounts for quantity purchases, ordering costs decrease while storage costs increase as a function of average inventory levels.* The determination of the economic order quantity requires the balancing of these two types of opposing costs.

For purposes of analysis and eventual development of an inventory model, therefore, one must identify all categories of costs, determine their relationship to various inventory levels, and measure them in relation to different strategies. Once the model has been developed, it can be solved by employing analytical, numerical, or simulation methods. This solution, in effect, yields the particular strategy that will result in an optimal value of the selected measure of overall effectiveness.

2.4.3 Queueing Models

Queueing, or *waiting-line*, *models* are designed to solve a class of problems in which a set of customers arrive for service at a set of service facilities. In such situations three things can happen. First, there is a perfect match between the requirements of the customers and the capacities of the service facilities. In such a case, we do not have a problem. Second, the demand for services exceeds the supply of services and the customers have to wait for service. Third, the supply of services exceeds the demand for services and hence the service facilities have to wait or remain idle. Obviously, specific costs are associated with the waiting of customers and the idle service facilities. Further, these two categories of costs move in the opposite directions.

For example, by adding more facilities we decrease the costs of customer waiting but increase the cost of idle facilities. The purpose of queueing models is to help design a system that will minimize a stated measure of performance, such as the sum of costs of customer waiting and idle facilities. Several management science models have been developed to solve queueing problems.

* With no quantity discounts and assumption of complete certainty regarding demand rate and lead time, the out-of-stock costs and costs of inventory items need not be considered by the model. Why?

2.4.4   Replacement Models

Customarily, *replacement models* are considered in two categories. The first deals with equipment that deteriorates over time. Equipment may lose efficiency as a result of use or the appearance of better equipment on the market. Lathes, drilling machines, planers, and electronic devices are representative of the type of equipment that deteriorates as a result of use or obsolescence. The second category deals with items that have a more-or-less constant efficiency over time, but when they fail, they do so suddenly and completely. A typical example of a problem in the second category is the decision as to a replacement policy for light bulbs.

Methods of analysis for formulating replacement policies with regard to these two categories are dissimilar because of the different nature and cost behavior of the equipment involved in each category. For equipment that deteriorates over time, for example, the analyst must consider, among other costs, the operating and maintenance costs that increase over time. It is necessary, when considering problems in this category, to decide upon an optimal interval of time after which the present equipment should be replaced with another candidate.

The category dealing with items that fail suddenly and completely calls for an analysis in which some sort of mortality distribution is predicted for the items in question. Based on this distribution the replacement policy has the objective of minimizing costs by determining a certain interval after which all items are replaced (within this interval individual items that fail are replaced immediately). Although this category of problems deals with decreasing, constant, and increasing probabilities of failure over time, a commonly used illustration is that of group replacement of light bulbs having an increasing probability of failure. In particular, the problem is to determine some interval of time such that the combined cost of replacing individual items within this interval and replacing all items at the end of the interval is minimized.

2.4.5   Maintenance Models

Maintenance models can be regarded as a special case of replacement or inventory models. They are replacement models in the sense that maintenance usually involves replacing parts after they fail to function effectively. They are inventory models to the extent that maintenance crews or facilities are stored to serve the maintenance needs of the future. Maintenance problems, like inventory problems, involve opposing sets of costs. For example, in a simple maintenance problem, one set results from machine or facility breakdown, and the magnitude of the costs in this set increases as the average idle machine time increases. Another set of costs results from measures adopted to decrease the average idle machine time, and these costs increase as the average idle machine time decreases. Thus the two sets of costs move in opposite directions as different levels of the average idle machine time are considered. The problem is to design and operate a maintenance program so that the sum of these two sets of costs is minimal.

2.4.6 Competitive Models

Competitive models help analyze those situations in which two or more *rational* opponents are involved in choosing strategies to optimize a measure of effectiveness. In the management literature, these models are included in the area of *game theory*. One category of game-theory models is that of two-person zero-sum games. In these games the number of players is two (hence two-person), and the loss of one player *exactly* equals the gain of the other (hence zero-sum). It is a measure of the general scope of linear programming that any two-person, zero-sum game can be formulated as a linear programming problem and then solved to obtain an optimal solution.

2.5 DECISION MAKING AND LINEAR PROGRAMMING

Linear programming, as mentioned earlier, is a decision model. It will be helpful to us in understanding the potential, as well as the limitations, of linear programming if we briefly examine various aspects of decision making, recognize the kinds of traditional and modern techniques that are available for decision making, and illustrate a linear programming problem with the aid of a sequence of steps always involved in a decision model.

2.5.1 What Is a Decision?

A *decision* is the conclusion of a process designed to weigh the relative utilities of a set of available alternatives so that the most preferred course of action can be selected for implementation. *Decision making* involves *all* the thinking and activities that are required to identify the most preferred choice. In particular, the making of a decision requires a set of goals and objectives, a system of priorities, an enumeration of alternative courses of feasible and viable actions, the projection of consequences associated with different alternatives, and a system of choice criteria by which the most preferred course is identified. As mentioned in Chapter 1, the essence of management is to make decisions that commit resources in the pursuit of organizational objectives. Planning, organizing, staffing, direction, control, leadership, communication, and all other functions of management are executed through the making and implementation of decisions.

2.5.2   Why Must Decisions Be Made?

Decisions must be made because we are living in a world in which resources are scarce, and because all human beings are motivated by a set of wants and needs. These needs can be biological, physical, financial, social, ego, or higher-level self-actualization needs. Each individual, in playing the various roles that he accepts in life, is motivated to act and make decisions in order to satisfy a set of wants and needs. We can say, in general, that people would act in such a manner as to create a state of affairs most conducive to satisfy their aspirations. These aspirations can be stated in terms of objectives and time-oriented goals. Decisions are made to achieve these goals and objectives.

2.5.3   Decisions and Conflict

One very obvious problem in decision making is to recognize the inherent conflict that exists among various goals relevant to any decision situation. For example, at the individual level, there is always the conflict between the desire to make more money and having more leisure time. Similarly, in the organizational context, there is always present the conflict of interest among various departments, divisions, and managers. The traditional conflict of interest among marketing, production, and financial departments can be described in terms of diversity of product lines, standard-ization, and requirements of working capital. The job of top management is to resolve such conflicts by making and implementing decisions.

2.5.4   Some Characteristics of Decisions and Decision Making

To appreciate the importance of decisions and hence the significance of decision theory, we note some of the salient characteristics of decisions and decision making. Decision making is a pervasive, deliberative, contin-uous, ubiquitous, and sequential activity. Its *pervasive* nature is obvious from the fact that all of us are engaged in the making and implementation of decisions in various roles that we accept in life. The *deliberative* nature of decision making is important because it implies an approach based on thought and reflection, as opposed to actions arising from habit and reflex. Decision making is a *continuous* as well as an *ubiquitous* process, because all human actions are related in one form or the other to the making and implementation of decisions. Although it is possible to analyze the making of a specific decision as an isolated phenomena, decisions are essentially *sequential* in nature. The sequential nature of decisions is obvious from the fact that in the interdependent world of today each decision has consequences and implications far beyond its original boundaries, drawn under simplifying assumptions.

2.5.5   Approaches to the Study of Decision Making

Decision theory, the body of knowledge that deals with the analysis and making of decisions, is an important area of study. This is evident from the significant contributions of such diverse disciplines as philosophy, economics, psychology, sociology, statistics, political science, and operations research to the area of decision theory. Of the several possible approaches to the study of decision making, we identify only two here. One relates to the preliminary questions of how to formulate goals and objectives, enumerate environmental constraints, identify alternative strategies, and project relevant payoffs. The second concentrates on the question of how to choose the optimal strategy when we are given a set of objectives, strategies, and payoffs. Linear programming has been applied essentially in the second approach—how to identify the optimal solution from a set of available alternatives.

Another way to classify the study of decision making is to consider the two categories of descriptive versus normative models. In the domain of *descriptive decision models,* the focus of study is on how people *do* behave and make decisions, not on how they *ought* to behave. The purpose here is to *describe* the process by which managers in fact go about making decisions. Behavioral science courses deal with this area of management.

Normative decision models, which are the main focus of economics and statistics, deal with how decisions *should* be made. These models prescribe for the manager the most preferred courses of action. The normative models (or prescriptive models) are so constructed that the decision criterion is part and parcel of the total model. All normative models are based on the assumption of rationality, which requires: (1) the ability to state objectives clearly and to rank them in some order of preference according to a set of decision criteria, and (2) the employment of proper means to optimize the achievement of objectives. Rationality also demands that when the time comes to choose a course of action, the manager makes the choice according to the *agreed-upon* criterion or criteria. Linear programming, as we mentioned earlier, is a prescriptive or normative model.*

2.5.6   Traditional and Modern Techniques for Decision Making

Table 2.1 shows some of the decision-making techniques that can be applied to solve programmed and nonprogrammed decisions. Most of the techniques listed in Table 2.1 are self-explanatory. The nature and purpose

---

* Whereas the normative descision models assume complete rationality and think of the manager as an "economic man" endowed with infinite wisdom, the descriptive decision models view the manager as a "administrative man" with limited knowledge and ability. The administrative-man idea is based on the assumption that in real life, managers do not really optimize and do not possess infinite rationality. Instead, the actual behavior of managers can be described more in terms of "satisficing" and "bounded rationality." See Simon [1957, p. 198].

Table 2.1. Traditional and Modern Techniques of Decision Making

	Decision-Making Techniques	
Types of Decisions	Traditional	Modern

Programmed
  Routine, repetitive
    decisions
  Organization develops
    specific processes for
    handling them
	1. Habit
	2. Clerical routine
	    Standard operating
	      procedures
	3. Organization structure:
	    Common expectations
	    System of subgoals
	    Well-defined infor-
	      mational channels
		1. Operations research
		    Mathematical
		      analysis
		    Models
		    Computer
		      simulation
		2. Electronic data
		    processing

Nonprogrammed
  One-shot, ill-structured,
    novel policy decisions
  Handled by general
    problem-solving
    processes
	1. Judgment, intuition, and
	    creativity
	2. Rules of thumb
	3. Selection and training
	    of executives
		Heuristic problem-
		  solving techniques
		  applied to:
		Training human
		  decision-makers
		Constructing heuristic
		  computer programs

Source: Simon [1960, p.8].

of simulation were described briefly in Section 2.3. Heuristic problem solving or heuristic programming is one of the categories of simulation models.* Basically, a *heuristic* is an aid to discovery or a rule of thumb to solve a particular problem; a *heuristic program* is a computer program that consists of a set of heuristics to be applied at various stages of the solution process.

2.6 BASIC STRUCTURE OF A DECISION MODEL

In each and every decision situation, regardless of its content or orientation, we propose a sequence of eight steps that must be executed, explicitly or implicitly. These steps constitute the basic structure of a decision model and are shown in Figure 2.4. We discuss next a sequence of eight steps of the basic decision model and illustrate them with respect to the product-mix example of Section 1.2. For quick reference, the technical specifications of the problem are reproduced in Table 2.2.

* Simulation models can be divided into four major categories: (1) Monte Carlo technique, (2) heuristic programming, (3) operational gaming, and (4) artificial intelligence. See Turban and Loomba [1976].

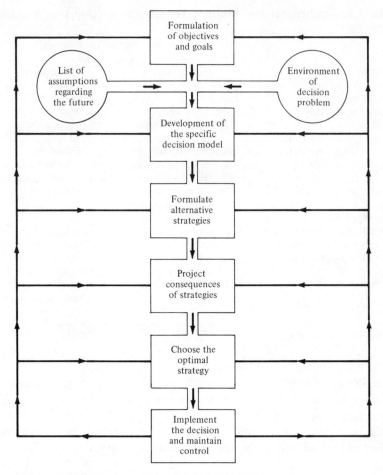

Figure 2.4 Basic Structure of a Decision Model

Table 2.2

	Product			Capacity Constraints per Time Period
Department	*A*	*B*	*C*	
Cutting	10	6	2	2,500
Folding	5	10	5	2,000
Packaging	1	2	2	500
Profit per unit	$23	$32	$18	

STEP 1. Formulation of a set of objectives and goals.

By the term *objective* we mean a qualitative statement of a desired occurrence or state of affairs. "The P/E ratio* of our stock should be improved" is an example of a statement of one specific objective. By the term *goal* we mean a time-oriented target. For example, "We should increase our share of the market by at least 10 per cent within the next year" is a statement of a goal. It would be useful for us to think of objectives and goals as being organized in an hierarchical system. That is, each objective can be subdivided into a set of goals. For example, the objective of increasing the P/E ratio can be pursued by attempting to achieve, say, these goals during the next fiscal year: (1) increase our share of the market in the Food Division by 10 per cent; (2) increase sales volume for the company by 30 per cent, either by expansion of the present markets or through acquisition; (3) initiate a companywide program to cut operating costs by 10 per cent; and (4) develop and test-market two new products.

A few comments are in order at this point. First, it is in the nature of things that objectives are usually multidimensional. Thus the manager must decide which dimension is relevant, what is an appropriate goal (time-oriented target) along each dimension, and what relative weights in terms of importance are to be attached to these goals. Second, if it is decided by the manager that, for purposes of analysis and planning, the objective is single-dimensional, the distinction between objective and the goal is only of theoretical value. Third, the manager must apply the appropriate scale for measurement along each dimension. Thus, although profits and costs are measured in dollars, it stands to reason that morale and motivation would have to be measured in nonmonetary units.

In the product-mix example of Section 1.2, the objective function required the maximization of profit: a single-dimensional objective to be measured in monetary units. (See also Table 2.2.) The linear objective function is given by: Maximize $23X + 32Y + 18Z$.

STEP 2. State the context and environment of the decision problem.

The purpose of this step is to "take stock" of the prevailing circumstances. In particular, the manager lists his resources; notes the constraints that delimit his freedom of action; and projects economic, political, and social trends, government policy (e.g., fiscal and monetary policy), and possible actions of his competitors. Then, a list of all the relevant variables that have a bearing on the problem is compiled and divided into two sets. One set consists of all those factors that are under the control of the manager (*controllable* factors); the second set consists of those factors over which the manager has little control (*noncontrollable* factors). The type and number of machines to be purchased, the number of shifts to operate, and the number of additional workers to be hired are examples of controllable factors. A

---

* *P/E ratio* refers to the ratio of stock price to earnings per share of stock.

particular combination of controllable factors is termed a *strategy*. A strategy is a course of action that the manager might employ to achieve his objectives.

The exchange rates among various currencies, the size of the federal budget, the number of strikes, and the number of new models that competitors will introduce next year are examples of noncontrollable factors. A specific combination of noncontrollable factors, emerging as a result of random and natural events, is called a *state of nature*.

The term *state of nature* is employed to describe a particular configuration of the noncontrollable factors that the manager will face as the future unfolds. When a manager lists one or more future states of nature, he is, in effect, projecting a specific set of events that will affect the outcomes of his decision. Thus a state of nature summarizes a configuration of noncontrollable factors emerging from several natural and random events, actions of rational opponents, or both.

In our product-mix problem of Section 1.2, and also in other linear programming problems, all variables are assumed to be controllable. Also, only a single state of nature is assumed, so that for every strategy there is a *unique* payoff. The environment of the decision problem is described by such assumptions as pure competition,* certainty, fixed technology, linearity, and divisibility.

It should be noted here that in real-life decision problems, the manager does not have complete freedom of choice even while selecting values for the controllable factors. The controllable factors can be manipulated only within certain ranges, and there are certain limits that cannot be violated. Such limits are made explicit by introducing a set of inequalities into the decision model.

As far as the noncontrollable factors are concerned, the manager attempts to study their behavior by building appropriate probability distributions. These distributions are then made the basis upon which forecasts regarding the behavior of the noncontrollable factors are made. For example, it is possible to record data and build appropriate daily probability distributions regarding the number of machine failures, number of man-hours required to repair them, and so on. These distributions can then be used to project future machine breakdowns, number of repairmen needed, and the like.

In linear programming, we assume complete control over the environment.

STEP 3. List assumptions regarding the future.

In this step, the manager can make different simplifying assumptions regarding the future. These assumptions have the impact of deciding what type of decision model will be employed. For example, the manager can assume conditions of complete certainty, or risk, or uncertainty. Further,

---

* Pure competition implies that price per unit remains constant during the planning horizon.

he could assume, albeit on the basis of statistical forecasts, specific values for some of the important factors, such as prime rate and tax rate. Similarly, assumptions can be made as to whether certain variables are to be treated as *continuous* or *discrete*, and whether the decision problem is to be treated as a *static* or as a *dynamic* problem.

Since the consequence, or payoff, of a decision is determined by the interaction of future circumstances with the chosen course of action, it is useful to classify the degree of future certainty or uncertainty that the manager is willing to assume. For our purposes, we shall consider three types of decisions, depending upon the assumed degree of certainty regarding the future: (1) decision making under complete certainty, (2) decision making under risk, and (3) decision making under complete uncertainty.* In the category of *complete certainty* we assume that the manager is capable of identifying all possible alternatives (or strategies) and that associated with each strategy is a *unique* outcome, payoff, or utility. The number of strategies in a given decision problem can be finite or infinite. Since decision making under certainty involves a situation in which each strategy is associated with a unique outcome, the problem for the manager is simply to compare all the outcomes, identify the *best* or *optimal* outcome, and then choose the cor-responding optimal strategy. Consider, for example, a machine shop in which three jobs are to be completed by utilizing three different machines. Assume that the time required to complete each job on each machine is known with complete certainty. The problem, then, is to assign these jobs to machines with the objective of minimizing production time. All the decision-maker has to do in such a case is to consider the various combina-tions in which jobs can be assigned and to calculate the production time associated with each combination. Of these, the decision-maker chooses the one with the minimum production time (or cost). The problem just described is the assignment problem presented in Section 1.9.4.

In the category of decision making under *risk*, there is more than one outcome (or payoff) associated with each strategy, but it is possible for the manager to attach a probability distribution to such outcomes. That is, the manager can state the probability with which a specific payoff will result, provided that a particular strategy is adopted. Consider, for example, that we decide to place a bet on the outcome of one flip of a coin. Obviously, there are only two possible states of the future: The coin will show either a head or a tail. Furthermore, in this case, it is possible to make an *a priori* statement about the probability of obtaining each future state. Similarly, one can state the probability of drawing an ace from a deck of cards, throwing 1, 2, . . ., on the roll of a die, or drawing a green chip from a bowl which contains, say, two green and three red chips, all on an a priori basis. In other cases, assuming conditions of stability and based on empirical results, probability statements

---

* Two other categories can be added to this classification: decision making under conflict, and decision making under partial information. For a discussion of these categories, see Miller and Starr [1969].

can be made in connection with the outcomes of some business events. For example, marketing research may indicate the respective probabilities of attaining different levels of future sales. In any case, the characteristic of a future involving risk is that the probability with which a particular state of affairs will occur can be stated, either on an a priori basis or on the strength of empirical data.

In the category of decision making under *complete uncertainty*, the manager projects more than one outcome (or payoff) for each strategy but is unable or unwilling to attach a probability distribution to such outcomes. In the final round of a table-tennis tournament, for example, we may sincerely refuse to project the chances of any one player's winning the match. Here there are two possible events for each player, win and lose, but because of insufficient supporting data, we may not be able to attach any objective probability values to the outcome. This simple illustration provides an example of an uncertain future. In the business world, managers attempt to gather relevant information so that some measure of probability can be attached to the various possible outcomes of a strategy. If this is possible, the uncertainty problem is converted to a decision problem under conditions of risk.

It should be obvious that linear programming falls under the category of decision making under complete certainty. The linear programming model is extremely useful because in many real-life problems it is economically desirable to *assume* complete certainty.

STEP 4. Development of the specific model.

The purpose of this step is to state, in an explicit fashion, the specific relationship among different variables and parameters that are relevant to the problem. What emerges is a specific model. The model could be a simple model that calls for optimization of a "reward," "payoff," or "utility" as a function of controllable or noncontrollable factors, as shown in

$$R = F(X_i, Y_j) \tag{2.1}$$

where

$R =$ the result, reward, or payoff measured in dollars, utility, or some other unit
$X_i =$ the set of controllable variables
$Y_j =$ the set of variables, and constants, which is not under the direct control of the manager but which affects R
$F =$ the functional relationship between the dependent variable R and the independent variables X_i and Y_j

For example, consider the inventory model given by Equation (2.2).

$$TC = \frac{D}{Q}S + \frac{Q}{2}CH \tag{2.2}$$

where

Q = order quantity (independent variable)

TC = total costs (dependent variable)

and D, S, C, and H are assumed to be known constants with respect to Q (i.e., their values do not vary as a function of Q).

In such models we can obtain *direct* solutions by using calculus.*

The model can also be of the variety in which something is to be maximized or minimized subject to a set of constraints (e.g., linear programming). Such models can be solved by an *iterative* approach; that is, several "passes" may be needed before the optimal solution can be identified. The simplex method of linear programming, to be presented in Chapter 7, represents an iterative approach.

STEP 5. Formulate alternative strategies.

The purpose of this step is to formulate alternative strategies that could be employed to achieve objectives. The specific model, developed in step 4, can be used to generate strategies.

Remember that we have defined the term *strategy* as a particular combination of controllable factors. The task of the manager here is to consider only those strategies that are feasible and viable in view of the available resources and specified constraints.

The degree of detail with which the alternative strategies are specified is determined by the nature of the problem, pressures of time, and the hierarchical level at which the decision is being made. Further, the number of strategies could be finite or infinite, depending essentially upon whether the decision variables are assumed to be discrete or continuous.

For example, in a typical linear programming problem, as represented in our product-mix example of Section 1.2, the number of feasible strategies is infinite. This is because the variables included in the model given by (1.4) are continuous. Any combination of products A, B, and C is a feasible strategy if it does not violate the constraints. Needless to say, we cannot list each strategy separately. What we need is an efficient method of search that will test only a finite number of strategies before an optimal solution is identified. The simplex method, to be presented in Chapter 7, is an efficient search procedure that will lead us, step by step, to successively better strategies,† until an optimal strategy is identified.

Similarly, the inventory model given by Equation (2.2) will yield an infinite number of strategies, provided that we again assume continuous variables. In some problems, however, reality dictates that only a finite number of strategies can be considered. Such problems, with a finite set of

---

* This is done by taking the first derivative of the total cost (TC) with respect to Q and then equating the first derivative to zero. The resulting equation yields the optimal value of Q.

† Technically, each new strategy can be as good as the preceding strategy, but it can never be worse. Normally, however, each new strategy is better than all previously tested strategies.

strategies, a finite number of states of nature, and hence a finite number of payoffs, can conveniently be expressed in the form of a matrix. Actually, even though the number of feasible solutions to a linear programming problem is infinite, the efficiency of the simplex method lies in the fact that it can identify the optimal solution in a finite number of steps.

STEP 6. Project consequences of alternative strategies.

Consequences of alternative strategies, or the payoffs associated with different strategies, are expressed in either monetary or utility terms. The mechanism through which such projections are made can vary from a sophisticated mathematical model to a simple intuitive prediction. For example, it is clear from the inventory model given in Equation (2.2) that for each specific value of the independent variable Q (order quantity), we shall have a unique value of the dependent variable TC (total cost). Similarly, in the linear programming model given in (1.4), each set of specific activity levels (i.e., each different production program) will yield a specific level of profit. At the very elementary level, especially where the number of finite strategies and the number of states of nature is very small, the consequences of alternative strategies may be predicted on an intuitive basis and shown in the form of a payoff matrix.

In linear programming the design of the simplex tableau (see Chapter 7) is such that the profit consequences of the specific strategy (or program) can be readily calculated.

STEP 7. Choose the optimal strategy.

This step requires that the optimal strategy be identified in accordance with a criterion of choice. Consider, for example, the general reward function given in Equation (2.1) and assume that we wish to maximize the value of R. This means that our criterion of choice is maximization, and we want to search for those values of X_i and Y_j that would give the maximum value of R. This task can be accomplished analytically, such as by calculus or by some iterative* procedure. Alternatively, we can employ simulation, a technique of problem solving that we mentioned in Section 2.3.

In the inventory-control model shown in Equation (2.2), the optimal strategy can be identified either by graphical means or by using calculus. In the linear programming problem, the optimal strategy is usually identified by the iterative approach (through the graphical, algebraic, or the simplex method).

STEP 8. Implement the decision and maintain a system of control.

This is the action step. Having identified the optimal strategy, the manager takes all the necessary steps to implement it. The question of follow-up and control, however, is of the utmost importance because the entire

---

* An *iterative* procedure is one in which the search starts from a specific point and takes us progressively to "better" points until, finally, the optimal strategy is identified.

decision model is based on a set of assumptions, including the assumption that the system is stable. If the optimal strategy is to remain optimal, a system of surveillance and control must be maintained to see whether any significant changes have taken place. If so, the circumstances would call for searching out a new optimal strategy. Depending upon the specific formulation of the decision problem, the decision environment, or the predilections of the manager, some of the eight decision steps may be combined or subsumed under different titles. But they must be considered and executed, explicitly or implicitly, consciously or unconsciously.

The idea of sensitivity analysis that we discuss briefly in Chapter 8 is helpful in maintaining a system of control in implementing optimal solutions designed through linear programming.

2.7 PROBLEM FORMULATION*

In Section 2.6, we discussed and illustrated a sequence of eight steps that must be executed, explicitly or implicitly, in each and every decision situation. The need for the decision arises because a problem is perceived or triggered by an information or control system. For example, an aggressive company president driven by ambition can arbitrarily set the goal of doubling his annual sales over a 5-year period. Starting with this goal, he will set in motion a series of actions (e.g., acquisition of other companies, introduction of new products, penetration of new markets) that lead to the need and desirability of defining or formulating problems at various levels of the organization. Similarly, a poor income statement can trigger the need to search for underlying causes that must then be formulated as problems. It is quite obvious that organizational problems are seldom isolated problems; they affect, and are in turn affected by, other problems in the organization. The determination of optimal inventory levels cannot be divorced from the consideration of available storage space or the constraints of working capital. However, in the real world, each problem *is, to some extent, isolated and then solved*. Its solution is then tested against the realities and power structure of organizational life. To a large extent, we proceed in exactly the same way when we formulate a linear programming problem. The problem is defined in terms of optimizing an objective function. Various solutions to the problem are then tested against the realities of available resources. The optimal solution is that which gives the best value of the objective function and does not violate any constraints. The search for the optimal solution can be conducted either manually or by using a computer. Thus, while formulating a linear programming problem, we must keep in mind these three aspects: (1) building the objective function, (2) building the constraints, and (3) preparation for solving the problem by computer.

* This section is adapted from Loomba and Turban [1974, pp. 58–59].

2.7.1 Building the Objective Function

The initial and probably the most difficult aspect of formulating a linear programming problem is the construction of its objective function. The function relates the objective or goal (dependent variable) to a set of independent variables and parameters. One of the first tasks is to identify both the independent variables and their structural relationship to the dependent variable. When such problems are formulated, our focus is usually on a single dependent variable, such as total profit, total cost, or percentage share of the market. Unfortunately, in many practical problems objectives are not single-dimensional; hence it is necessary to convert a multiple-objective problem to a single-objective problem. We now describe three approaches for handling multiple-dimensional problems.

Expressing All Objectives But One in the Form of Constraints

Suppose that a company has three objectives: maximizing profit, maximizing the share of the market, and maximizing the growth rate. It could then specify two of these objectives in the form of constraints, thus leaving a single objective function to be maximized. For example, the company can state its objective as the maximization of its share of the market, subject to the constraints that it achieve at least $1.50 profit per share and at least a 10 per cent annual rate of growth. An important consideration in this approach is how to choose those objectives that should be expressed as constraints and then determine their minimum acceptable levels.

Expressing All Objectives as a Single Objective

Theoretically, it is possible to express any number of objectives as a single objective if we can find a *common denominator*, such as money or utility, that will express the trade-off relationships among the various objectives. The major problem in such an approach is how to build the trade-off relationships among the various objectives.

In many practical cases one finds that when multiple objectives are expressed as a single objective function, the resultant objective function is often nonlinear. However, nonlinear models are usually more difficult to solve than linear programming problems. The reason for considering the question of multiple objectives rather than a single objective is to bring the model as close to reality as possible. However, the attempt to reflect reality can often lead to difficulty, both in terms of formulating the problem and building its mathematical model. The result is that in several cases the models are so complicated that it is impossible (or too expensive) to find optimal solutions for them. The only recourse, in such cases is either to sacrifice part of the reality and arrive at a simpler formulation, or to seek "approximate" or "good enough" (rather than optimal) solutions for the more realistic but complicated problem. Finding the proper balance between the reality and the amount of complexity is a major problem for the model builder.

Goal Programming*

Goal programming is another method of handling multidimensional problems. In goal programming, we list various goals. rank them in terms of priority, and specify quantitative targets for each goal. These targets are incorporated as constraints, and the objective function is made up of deviations (overachievement as well as underachievement) from specified targets. The objective in goal programming is the minimization of these deviations. It should be noted that the assumed behavior in goal programming is that of a *satisficer* rather than of an *optimizer*.

2.7.2   Building the Constraints

The specification of the constraints is derived from the existing state of technology; available levels of physical, economic, and other resources; and stated requirements relating to important factors. Each constraint is built after estimating the respective input–output coefficients that tie a resource, or a requirement, to specified decision variables (activities). Wherever possible, only the significant factors should be expressed as constraints, since the computational work in linear programming is proportional approximately to the cube of the number of constraints involved, and since an excessive number of constraints might outstrip the capabilities of the computer.

2.7.3   Preparation for a Solution by Computers

Most real-life problems are too large and complex to be solved manually, and the availability of a computer, therefore, is a necessity. In formulating and solving a linear programming problem, we should bear in mind that the problem should be so structured that it will be possible to solve it by computers.

A large number of computer packages (programs) are available that can be used to solve many problems if the problems are structured to conform to the package design. In some cases, all we have to do is to provide the computer with coefficients of the objective function, the input–output coefficients of the constraints, and the number of variables; the output is the solution to that problem. However, if the problem cannot be simplified to fit one of the packages, a program has to be devised to meet its particular requirements. This can be a costly and time-consuming process.

* For a very lucid explanation of goal programming, see Lee [1974].

Computers are a very important but not the only aspect of linear programming applications and implementation. We shall discuss this matter further in Chapter 12.

2.8 SUMMARY

In this chapter we have presented various apsects of two important concepts, models and decision making. The term *model* was defined, the purpose of models was explained, and the concept of model building was briefly discussed. Different classifications of models were presented so that the characteristics and assumptions of the linear programming model can be related to a specific category of models. In this manner, the place, potential, and limitations of the linear programming model can be more clearly grasped. For the same purpose, a brief discussion of six classes of management science models was presented. Since linear programming is a decision model, we examined several aspects of *decision making*. Various approaches (e.g., normative and descriptive) to the study of decision making were mentioned and their relationship to linear programming indicated. A table containing a list of traditional and modern tools for decision making was provided. Linear programming is a modern tool that can be most effectively used in solving repetitive decisions with specified structures.

We then presented a basic decision model and differentiated between direct and iterative approaches to the solution of analytical models. Finally, and most important, a sequence of eight steps of a decision model was listed and explained with reference to a linear programming problem.

The overall purpose of the first two chapters was to give a broad picture of how linear programming relates to the concepts of management, models, and decision making. In a sense, we have provided, in the first two chapters, a macro orientation from which the reader can appreciate the rest of the book. We now propose to concentrate on the general model of linear programming, and the various approaches of analyzing and solving linear programming problems. Chapter 3 is devoted to a discussion of the graphical approach.

REFERENCES

Dantzig, G. B. *Linear Programming and Extensions*. Princeton, N.J.: Princeton University Press, 1963.
Forrester, J. W. *Industrial Dynamics*. Cambridge, Mass.: The MIT Press, 1961.
Lee, S. M. *Goal Programming for Decision Analysis*. Philadelphia: Auerbach Publishers, Inc., 1974.
Levey, S., and N. P. Loomba. *Health Care Administration*. Philadelphia: J. B. Lippincott Company, 1973.

Loomba, N. P., and E. Turban. *Applied Programming for Management*. New York: Holt, Rinehart and Winston, Inc., 1974.

Miller, D. W., and M. K. Starr. *Executive Decisions and Operations Research*, 2nd ed. Englewood Cliffs, N.J.: Prentice-Hall, Inc., 1969.

Simon, H. A. *Models of Man*. New York: John Wiley & Sons, Inc., 1957.

Simon, H. A. *The New Science of Management Decision*. New York: Harper & Row, Publishers, 1960.

Turban, E., and N. P. Loomba. *Readings in Management Science*. Dallas, Tex.: Business Publications, Inc., 1976.

REVIEW QUESTIONS AND PROBLEMS

2.1. What is the purpose of building models? To be useful, what characteristics must a model possess?

2.2. List various classifications of models. Describe the type of problem that can be solved by different models under each classification. Describe an actual business problem and suggest a model to solve it.

2.3. Distinguish between prescriptive and descriptive models. Are the following models prescriptive or descriptive?

(a) Linear programming.

(b) Balance sheet.

(c) Tax returns.

(d) Annual budget.

(e) Opinion polls.

2.4. Describe static and dynamic models. Give two examples of each type of model.

2.5. What types of systems can best be studied through the use of simulation models? Describe the essential characteristics of simulation models.

2.6. Describe the form of the relevant problem for which the following management science models can be used.

(a) Allocation models.

(b) Inventory models.

(c) Replacement models.

2.7. Discuss various approaches to decision making, with particular reference to normative and descriptive decision theories.

2.8. Briefly explain programmed and nonprogrammed decisions. Illustrate each by real-life problems. What problem-solving techniques are used in each of the two categories?

2.9. What are the basic steps involved in making a decision? Choose a problem situation and describe how by the application of these steps a better decision can be made.

2.10. Describe the important phases of problem formulation as discussed in this chapter.

3
Linear Programming: A Graphical Approach

3.1 INTRODUCTION

As stated in Chapter 1, linear programming is a method of allocating limited resources among two or more interdependent activities or competing candidates. The allocation of resources is determined with the objective of maximizing or minimizing a linear objective function that represents some criterion of effectiveness, such as profit, cost, or sales. How is this allocation determined? What are the different methods for handling such problems? What are the actual mechanics of these methods? We shall answer these questions by examining a typical linear programming problem and solving the *same* problem by employing different methods of solution. In particular, we shall analyze and solve our problem by (1) the graphical method, (2) the algebraic method, (3) the vector method, and (4) the simplex method. The purpose of concentrating on the same problem is to enable the reader to grasp the relationships among the different solution stages involved in the various methods for solving linear programming problems.

Of the above-mentioned approaches to linear programming, the simplex method is the most general and powerful. The algebraic and the vector approaches are, in a sense, the foundations on which the simplex method is built. Each method, when and if applied to a given problem, will lead to the same optimal solution. Each provides a different perspective of how a series of systematic steps leads from one solution to a "better" solution, and finally to the "best" solution. We present the algebraic approach in Chapter 4.

The vector approach and the simplex method are discussed in Chapters 6 and 7. The purpose of this chapter is to examine linear programming from a graphical perspective.

When stated mathematically, the structural constraints of any linear programming problem consist of a certain number of rows and a certain number of columns. The number of rows is determined by the number of resources, or characteristics, specified by the technology of the problem. For example, in the product-mix problem of Section 1.2 we have three rows. The number of columns is determined by the number of activities or candidates. For example, in the example of Section 1.2, we have three columns. In this book we shall always denote the number of rows by the letter m and the number of columns by the letter n. The number of rows and the number of columns of the structural constraints determine the dimensions of a linear programming problem. In general, we can think of a linear programming problem as an $m \times n$ (read m by n) problem.

Let us examine again the problem stated by (1.4), whose structural constraints set has 3 rows and 3 columns. This can be viewed as a 3×3 (read 3 by 3) problem, involving 3 resources and 3 activities. Similarly, the diet problem given in Table 1.3 is a 2×2 problem, involving 2 requirements and 2 activities. How can we proceed to graph such problems of $m \times n$ dimensions?* One obvious restriction is that in order to graph a set of structural constraints, we are limited to those cases in which either the number of rows, or the number of columns, is 2 or less.† If the number of columns (i.e., activities) is 2, we can consider any $m \times 2$ problem and graph the constraints in the two-dimensional *activity* space. For example, we have graphed in Figure 3.6 the problem that is shown in Table 3.2 and stated mathematically by (3.1). If the number of rows (i.e., resources) is 2, we can consider any $2 \times n$ problem and graph the constraints in the two-dimensional *resource* space. For example, we have graphed in Figure 3.15 the problem that is shown in Table 3.4 and stated mathematically by (3.10).

The preceding remarks were made to indicate that, theoretically, the graphical approach can be employed to solve any $m \times 2$ or $2 \times n$ linear programming problem. That is, we are, for all practical purposes, restricted to a two-dimensional space. The graphical method is, therefore, not an efficient method of solving large real-life linear programming problems. The value of the graphical approach lies in its ability to show, without the benefit of a rigorous mathematical proof, several different characteristics of linear programming problems and their solutions. Furthermore, familiarity with the graphical method can be helpful in understanding the mechanics and rationale of other methods of solving linear programming problems that we present in subsequent chapters.

* The quantities m and n can each equal 1, 2, 3, ...
† It is possible to graph in a three-dimensional space; however, we ignore that option because it is neither efficient nor does it have any significant pedagogical value.

In this chapter we present the graphical approach in terms of activity space as well as the resource space. From such an analysis we shall be able to arrive at several useful generalizations that will give us a "feel" of linear programming.

3.2 THE PROBLEM

By means of some sharp bargaining with the union and the subsequent reduction of union "make-work" restrictions in his former contract, a small paper-towel manufacturer has created additional capacity in each of his three main production departments: cutting, folding, and packaging. For purposes of identification, three sizes of paper towels currently produced by the company will be called products A, B, and C. Owing to its small size, the company can sell in the market all that it can produce at a constant price. Management is inclined to be conservative and does not wish to expand production facilities at this time, although they do wish to utilize as much of the recently created additional capacity as is possible.

The paper toweling is received from another manufacturer in the form of large rolls. These rolls are subsequently cut, folded, and packaged in the three sizes. The pertinent manufacturing and profit information for each size of paper towel is summarized in Table 3.1.

Table 3.1

Department	Product A	B	C	Capacity Constraint per Time Period
Cutting	10	6	2	2,500
Folding	5	10	5	2,000
Packaging	1	2	2	500
Profit per unit	$23	$32	$18	

The data in the table represents the technical specifications of the three products. The production of 1 unit of size A, for example, requires 10 units of processing time (say minutes) in the cutting department, 5 minutes in the folding department, and 1 minute in the packaging department. When sold, product A yields a profit contribution of $23 per unit. The total available capacity in the cutting department is 2,500 minutes. Similar information is available in Table 3.1 for producing 1 unit each of products B and C. Note that each product must be processed through all three departments.

A quick examination of the data in Table 3.1 also reveals that product A is more efficient than product B in the use of folding capacity, whereas product B is more efficient in its demand on cutting capacity. Similar observations can be made about product C. Different degrees of efficiency of products competing for the use of various available resources are typical in linear programming problems.

Our problem is to determine the optimal program of production. That is, we must determine the mix of products that will maximize profits by utilizing the additional capacity created by the new union contract. We repeat here the assumptions of linearity and continuous variables.

3.3 THE GRAPHICAL METHOD (ACTIVITY SPACE)

First we present a *maximization* case. The problem shown in Table 3.1 is a 3×3 problem. For the purpose of illustrating the graphical approach, let us assume that our manufacturing company has arbitrarily decided to produce only products A and B. The pertinent data for this modified problem are given in Table 3.2.

Table 3.2

| | Product | | Capacity per |
Department	A	B	Time Period
Cutting	10	6	2,500
Folding	5	10	2,000
Packaging	1	2	500
Profit per unit	$23	$32	

Our objective is to determine the *mix* of products A and B that will yield the maximum profit. The problem can be stated mathematically as follows:

Maximize $\qquad\qquad 23X + 32Y$

subject to the constraints

$$10X + 6Y \leqslant 2,500$$
$$5X + 10Y \leqslant 2,000 \qquad\qquad (3.1)$$
$$1X + 2Y \leqslant 500$$

and

$$X, \quad Y \geqslant 0$$

The graphical approach consists of graphing the structural constraints and thereby determining the region of feasible solutions, graphing the objective function, and identifying the optimal solution.

3.3.1 Graphing the Structural Constraints

As mentioned in Chapter 1, the structural constraints can be of three types: (1) equality ($=$), (2) less than or equal to ($\leqslant$), or (3) greater than or equal to ($\geqslant$). Next, we provide an example of each type:

Constraint	Graphical Representation
$2X + 4Y \leqslant 20$	Figure 3.1a
$5X + 3Y \geqslant 30$	Figure 3.1b
$2X + 4Y = 20$	Figure 3.1c

Note that in Figure 3.1a, the "less than or equal to" type of constraint is represented by the entire shaded area below the line $2X + 4Y = 20$. Similarly, in Figure 3.1b the "greater than or equal to" type of constraint is represented by the entire shaded area above the line $5X + 3Y = 30$. In Figure 3.1c, the equality constraint is represented by a line alone; no shaded area is involved. It should be noted here that if we plot the constraints $X \geqslant 0$ and $Y \geqslant 0$ in each of Figures 3.1a through 3.1c, the result will be to restrict everything to the first quadrant. Let us now return to our problem of Table 3.2.

It is clear that our first task is to state the structural constraints of the problem. For each of the three constraints embodied in Table 3.2 we can write an inequality* of the "less than or equal to" type. It is desirable that the reader gain familiarity with these inequalities and their physical interpretation. Let us consider the inequality that represents the cutting department capacity. If we let variables X and Y denote, respectively, the units of products A and B to be produced, the inequality for the cutting department is

$$10X + 6Y \leqslant 2{,}500$$

This inequality is simply an algebraic way of expressing the information given in Table 3.2 in connection with the cutting department. It gives algebraic expression to one of the three structural constraints in our problem. A descriptive translation of this inequality is: Each unit of product A requires 10 units of the cutting capacity, and each unit of product B requires 6 units of the cutting capacity; the amount of A (i.e., a specific value of X) and the amount of B (i.e., a specific value of Y) to be produced should be such

---

* For a discussion of inequalities, see Section 4.2.

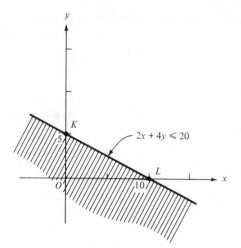

Figure 3.1*a*

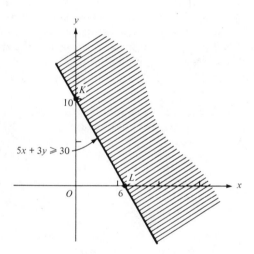

Figure 3.1*b*

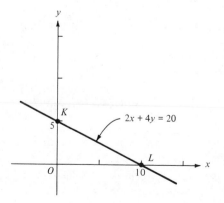

Figure 3.1*c*

that the total demand on the capacity of the cutting department does not exceed 2,500 units. Thus any specific combination of values of X and Y that does not violate this and the other given constraints is a *feasible* solution.

Let us assume that we wish to produce just 2 units of product A and 2 units of product B. This program ($X = 2$, $Y = 2$) can give us two types of information in connection with each resource: (1) the total capacity used by this program, and (2) whether or not the capacity constraint has been violated. For this program, the total cutting capacity used is $10(2) + 6(2) = 32$ units, and the remaining cutting capacity, therefore, is $2,500 - 32 = 2,468$ units. The constraint on the capacity of the cutting department obviously has not been violated. For each program, a similar check must be made of all resources.

As can easily be ascertained, $X = 2$ and $Y = 10$ is also a possible solution for this problem. Furthermore, this program results in a higher level of profit than our first program of $X = 2$ and $Y = 2$. However, there may be other programs that will yield even larger profits than the second program ($X = 2$, $Y = 10$). If so, we must discover them. This suggests that the search for a better program must continue until an optimal solution is determined. The optimal solution for our problem is the specific program that will yield the highest level of profit without violating any of the structural constraints. We shall identify the optimal solution by employing the graphical approach.

The first task in the graphical approach is to write all the constraints for the problem and then graph them in the activity space.* In our problem, for cutting capacity

$$10X + 6Y \leqslant 2,500 \tag{3.2}$$

for folding capacity

$$5X + 10Y \leqslant 2,000 \tag{3.3}$$

and for packaging capacity

$$1X + 2Y \leqslant 500 \tag{3.4}$$

This set of inequalities can be graphed easily. Inequality (3.2), for example, is graphed in Figure 3.2.

To obtain the X and Y intercepts for inequality (3.2), we proceed as follows: Let $X = 0$; then

$$Y = \tfrac{2,500}{6} = 416\tfrac{2}{3} \qquad \text{(point } K\text{)}$$

Let $Y = 0$; then

$$X = \tfrac{2,500}{10} = 250 \qquad \text{(point } L\text{)}$$

---

* We speak of *activity space* because here the two-dimensional space in which we graph the inequalities will have as its coordinates the activities X and Y.

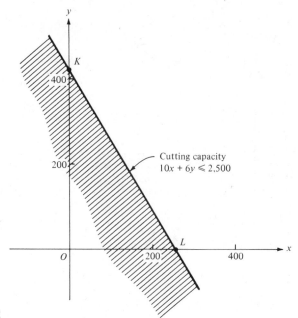

Figure 3.2

Joining points K and L gives us a line whose equation is

$$10X + 6Y = 2,500$$

However, since we do not wish to plot the above equation, but rather the inequality $10X + 6Y \leqslant 2,500$, the region of interest is represented by the shaded area in Figure 3.2. Clearly, the shaded area also includes negative values of X and Y which imply negative production and have no real physical counterpart. In order to exclude any possibility of negative production, a set of nonnegativity constraints is introduced. In our example, the nonnegativity constraints are $X \geqslant 0$, $Y \geqslant 0$; and they imply that we are restricted to producing zero or more units of products A and B.

The addition of nonnegativity constraints restricts the area of possible solutions to the first quadrant of the XY plane, as shown in Figure 3.3.

Similarly, Figures 3.4 and 3.5 show the areas of possible nonnegative solutions for the folding and packaging capacities, respectively.

If we combine Figures 3.3, 3.4, and 3.5, we obtain Figure 3.6, the shaded area of which represents the region of all feasible solutions for our problem.

Any point in the shaded area and/or on the boundary of the shaded area of Figure 3.6 is a possible solution. In other words, there are an infinite number of solutions for this problem if we assume divisibility of the production units. Our objective is to pick at least one point (X, Y) from the shaded (feasible) area of Figure 3.6 that will maximize profit or yield the highest value for the linear objective function: $23X + 32Y$.

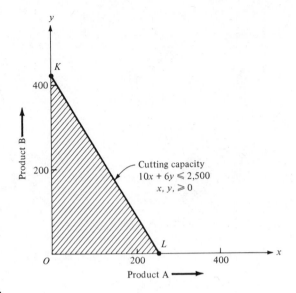

Figure 3.3

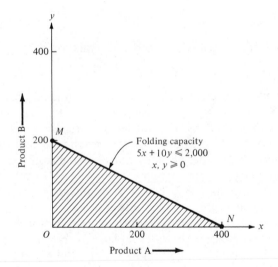

Figure 3.4

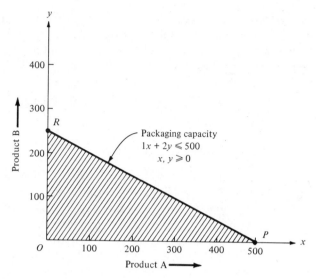

Figure 3.5

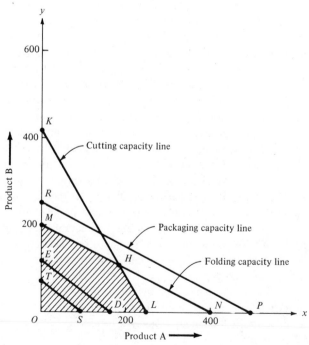

Figure 3.6

How can we proceed to accomplish this? It is at this point that we are guided by the objective function. If, somehow, we can graph the objective function in Figure 3.6, determine its direction of maximum increase, and start and keep on moving it in this direction, it will eventually touch some farthest point on one of the boundary lines of the shaded area. This point, then, will give us a unique (or one of many possible) optimal solution.

3.3.2 Graphing the Objective Function

The objective function can be of the minimizing or the maximizing type. For our problem we are to maximize the linear profit function: $23X + 32Y$. This function can be graphed for different values of total profit, such as $736 or $1,472. The graph of the linear profit function for a specific amount of profit results in a straight line and is known as an *isoprofit line*. Each point on a specific isoprofit line corresponds to a "program"; and each point yields the same profit as any other point on that line. How do we graph an isoprofit line? By choosing a specific profit level, finding those values of X and Y that yield this profit level, and then joining these two values with a straight line. For example, how many units of product A alone, and how many units of product B alone, are required to produce a profit of, say, $736.* Since profit per unit of product A is $23, the answer is obviously $X = 32$. Similarly, since profit per unit of product B is $32, it will require a value of $Y = 23$ to produce a profit of $736. Thus one set of two isoprofit points is $X = 32$, $Y = 0$; and $X = 0$, $Y = 23$. Another set of isoprofit points in Figure 3.6 is $X = 96$, $Y = 0$ (point S), and $X = 0$, $Y = 69$ (point T). When we join these two points (S and T), we obtain the $2,208 isoprofit line ST, as shown in Figure 3.6. All points on this line are within the region of feasible solutions; hence all represent feasible programs, each yielding a profit of $2,208. Thus an isoprofit line is the locus of all points (i.e., all possible combinations of X and Y) that yield the same profit. In a similar fashion we could have drawn other isoprofit lines, yielding different levels of profit contribution. For example, line DE in Figure 3.6 represents the $3,680 isoprofit line. A comparison of lines DE and ST shows that the $3,680 isoprofit line ($DE$) is parallel to the $2,208 isoprofit line ($ST$) (i.e., they have the *same* slopes) and is located farther away from the origin. This was to be expected, since the per unit profit of each product is fixed, and larger total profit will be obtained as we move away from the origin. Note also that, in the two-dimensional activity space, the slope of any isoprofit line is determined by the profit coefficients in the linear objective function. For example, in our problem, the slope of any profit line is given by $-\frac{23}{32}$.

---

* Any arbitrary profit figure will work. However, it is always convenient to pick a profit number that gives an integer as an answer and places the isoprofit line within the region of feasible solutions.

3.3.3   Finding the Optimal Solution

Let us examine the convex* polygon $OMHL$ shown in Figure 3.6. We assert, without giving a formal mathematical proof, that an optimal solution to a linear programming problem can always be found at one of the corner (or extreme) points of the region of feasible solutions. This can be seen by examining three possible situations regarding the relationships between the linear objective function of a problem involving just two activities. Let us examine the feasible region $OMHL$ in Figure 3.6 which reflects our two-activity problem. First, if per unit profit of A is overwhelmingly large as compared to per unit profit of B, we must produce A alone, and hence the optimal solution must be at the corner point L. Second, if the per unit profit of B is overwhelmingly large as compared to per unit profit of A, we must produce B alone and the optimal solution must be at the corner point M. Third, if the relationships of per unit profits for products A and B is such that the optimal solution cannot include just one product, but must include both products, the optimal solution *can* always be located at a corner point, such as H in Figure 3.6. The truth of the third case can easily be

---

* A convex polygon consists of a set of points having the property that the segment joining any two points in the set is entirely in the convex set. There is a mathematical theorem which states: *The points that are simultaneous solutions of a system of linear inequalities of the "less than or equal to" type form a polygonal convex set.* For purposes of visualization refer to Figures 3.7 and 3.8; the former represents a polygonal convex set, whereas the latter does not. Note that if points C and D in Figure 3.8 are joined, the definition of a convex polygon is seen to be violated.

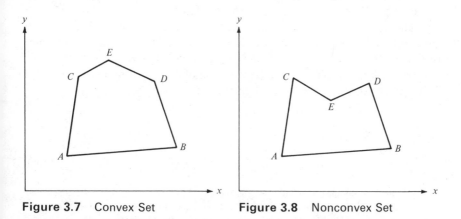

Figure 3.7   Convex Set   **Figure 3.8   Nonconvex Set**

Although Figures 3.7 and 3.8 are in two-dimensional space, the concept of a convex set is general and can be extended to any n-dimensional space. Furthermore, since all linear programming problems contain structural constraints that are of the "less than or equal to" type or, if not, can be converted to the "less than or equal to" type, the solutions to linear programming problems form a convex set.

substantiated if we consider the slope of an isoprofit line with respect to the slopes of lines MH and HL in Figure 3.6.*

The above remarks were made with reference to Figure 3.6 but are meant to assert the following general statement: *If there exists a solution to a linear programming problem, then an optimal solution can always be found at an extreme (or corner) point of the convex set of feasible solutions.*

Let us hark back to our problem and the consideration of the isoprofit lines in Figure 3.6. Since the objective is to maximize profit, we should keep on drawing such isoprofit lines for higher profits as long as any part of the highest isoprofit line remains within the feasible (shaded) region. Obviously, we shall have to stop only when we have hit either a corner point of the convex polygon $OMHL$ or one of its boundary lines. In either case we would have found our optimal solution(s).

In a profit-maximization problem involving only two competing candidates, the isoprofit line farthest from the origin but still within the feasible-solution area is used to determine the optimal solution. Two cases can arise. First, this farthest feasible isoprofit line may be coincident with one of the boundary lines of the convex polygon. If this is the case, all points on the boundary line that is coincident with the isoprofit line are feasible, as well as optimal, solutions.† Second, the farthest feasible isoprofit line may not coincide with any of the boundary lines of the convex polygon. If this is the case, one of the corner points of the convex polygon provides the optimal and unique solution.

In our case, we observe that the isoprofit line farthest from the origin and still within the region of feasible solutions passes through the point H. Hence point H represents the optimal solution from among an infinite number of solutions represented by the shaded area of Figure 3.6.

The coordinates of the point H can be determined either directly from the graph or by a simultaneous solution of the two lines intersecting at point H. The equations of these lines, representing the cutting and folding capacities, are already known to us. The determination of the coordinates of point H is illustrated next. For the cutting department:

$$10X + 6Y = 2,500$$

or

$$Y = \tfrac{1,250}{3} - \tfrac{5}{3}X \qquad\qquad (3.5)$$

---

* If the isoprofit line is parallel to MH, each point on MH (including point H) is an optimal solution. Similarly, if the isoprofit line is parallel to HL, each point on HL (including point H) is an optimal solution. If the isoprofit line is not parallel to any boundary line (such as MH and HL), it pays to move it as far away from the origin as possible, and we shall eventually stop at point H.

† In a two-candidate case, if the slope of any of the boundary lines is the same as that of any isoprofit line, we can conclude that the linear programming problem has many alternative optimal solutions. In such a case any isoprofit line will be either parallel to or coincident with one of the boundary lines of the convex polygon.

For the folding department:

$$5X + 10Y = 2,000$$

or

$$Y = 200 - \tfrac{1}{2}X \tag{3.6}$$

Equating Equations (3.5) and (3.6), we obtain

$$\tfrac{1,250}{3} - \tfrac{5}{3}X = 200 - \tfrac{1}{2}X$$

or

$$X = \tfrac{1,300}{7} \tag{3.7}$$

Substituting Equation (3.7) in (3.6), we find that

$$Y = 200 - \tfrac{1}{2}(\tfrac{1,300}{7})$$

$$= \tfrac{750}{7}$$

Hence the optimal solution is to produce $\tfrac{1,300}{7}$ units of X (product A) and $\tfrac{750}{7}$ units of Y (product B). The profit for this program is

$$23\tfrac{1,300}{7} + 32\tfrac{750}{7} = \$7,700$$

When we substitute $X = \tfrac{1,300}{7}$ and $Y = \tfrac{750}{7}$ in the inequalities (3.2) to (3.4), we note that this program fully utilizes the capacities of the cutting and folding departments but leaves 100 units of the packaging department capacity unused. This can also be observed by examining Figure 3.6, in which the packaging-capacity line is far above the shaded area.

Referring again to Figure 3.6, we make the following comments:

1. Any solution that does not violate the structural and the non-negativity constrained is a feasible solution.
2. Because of the assumption that all variables are continuous, a linear programming problem has an infinite number of solutions. Each and every point in the set $OMHL$ represents a possible or feasible solution. The set of such solutions is a convex set, regardless of the number of activities or constraints involved. For the two-dimensional case, the convex set is represented as a shaded area such as that shown in Figure 3.6.
3. The structural constraints can be of two types: redundant or non-binding, and active or binding. As can be seen in Figure 3.6, the packaging-capacity constraint is nonbinding. That is why the line representing the packaging capacity does not determine any of the boundary lines of the convex set $OMHL$. Also, because the packaging capacity is nonbinding, it is not fully utilized. Note that our optimal solution leaves 100 units of the packaging capacity unused.
4. Of the infinite number of solutions defined by the convex set, the optimal solution can always be found at its extreme points. Hence an

efficient search technique is to test only the extreme point (or corner) solutions for optimality. Any extreme point solution is called a *basic feasible solution.** The solutions at points O, M, H, and L are basic feasible solutions.

3.4 THE MINIMIZATION PROBLEM

In this section we shall apply the graphical method to solve a minimization problem. The structural constraints of the problem will be graphed in the activity space.

Table 3.3

Nutrient	Food		Weekly Requirement
	A	B	
I	2	3	3,500
II	6	2	7,000
Cost per unit of food (cents)	10.0	4.5	

Let us consider a minimization problem in which we are to determine the least-cost diet. The problem calls for meeting minimum daily requirements of two nutrients by purchasing a mix of two foods. The specifications for the problem are shown in Table 3.3. The problem can be expressed mathematically as:

Minimize $\qquad\qquad 10X + 4.5Y$

subject to the constraints

$$2X + 3Y \geqslant 3,500$$
$$6X + 2Y \geqslant 7,000 \qquad\qquad (3.8)$$

and

$$X, \qquad Y \geqslant 0$$

We have graphed this problem in Figure 3.9.

* Actually, there are two types of basic feasible solutions: *nondegenerate basic feasible* and *degenerate basic feasible*. The distinction is in terms of number of positive variables in a solution and the number of structural constraints, m. If the number of positive variables in the solution equals m, we have a nondegenerate basic feasible solution. If it is less than m, we have a degenerate basic feasible solution. In this book, we shall use the term "basic feasible solution" or just "basic" solution to denote nondegenerate basic feasible solutions. A degenerate basic feasible solution is shown by point A in Figure 3.11. Also see Figure 5.7.

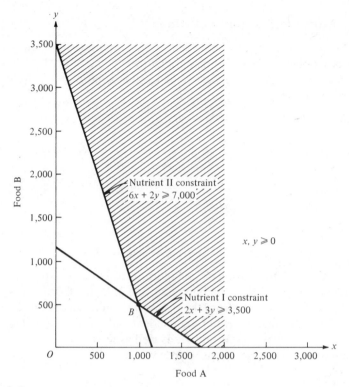

Figure 3.9

The graphical procedure for solving a two-activity minimization case is exactly the same as the one described for the maximization case. The only difference is that the linear objective function is graphed in the form of *isocost lines*, as opposed to the isoprofit lines of the maximization case. After the structural constraints of the problem have been graphed to define the region of feasible solutions, we can draw isocost lines of successively lower values. The isocost line that is closest to the origin and yet lies within the convex set will help identify the optimal solution. The optimal solution, as the reader can verify from Figure 3.9, is found at the corner point B.*

3.5 SPECIAL CASES

In most cases, linear programming problems yield unique optimal solutions. However, the following special cases should be noted: (1) multiple optimal solutions, (2) degenerate basic feasible solution, (3) unbounded solution, and (4) no feasible solution.

* The optimal solution: food $A = 1,000$ units, food $B = 500$ units, and total cost $= $122.50.

3.5.1 Multiple Optimal Solutions

A linear programming problem yields multiple optimal solutions whenever its objective function is parallel to any one of the boundary lines (i.e., constraints) that define the set of feasible solutions. Let us illustrate by considering the following problem.

Maximize $10X + 20Y$

subject to the constraints

$$10X + 6Y \leqslant 2{,}500$$
$$5X + 10Y \leqslant 2{,}000$$
$$1X + 2Y \leqslant 500$$

and

$$X, \quad Y \geqslant 0$$

We have graphed this problem in Figure 3.10. Note that the $2,000 isoprofit line FG is parallel to the boundary line MH. If we draw successively higher isoprofit lines away from the origin, we can observe that the highest isoprofit line is the one that will be coincident with the line MH. This means that the two corner points M and H, as well as any other point on the line MH, represent optimal solutions. Hence we have a special case in which we

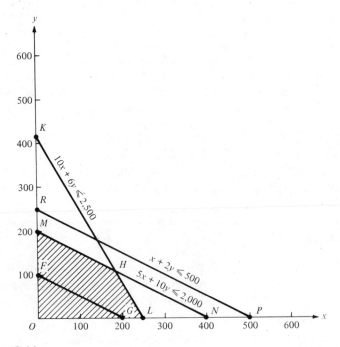

Figure 3.10

have multiple optimal solutions rather than a unique optimal solution.*
The existence of multiple optimal solutions gives the manager some degree of
flexibility in terms of determining the exact product mix.

3.5.2   The Degenerate Basic Feasible Solution

A degenerate basic feasible solution, like the nondegenerate basic
feasible solution, is also a corner point solution. However, the degenerate
basic feasible solution has the property that the number of positive variables
that comprise the solution is *less* than the number of structural constraints
m. A graphical illustration of a degenerate basic feasible solution is provided
in Figure 3.11.

Let us consider the following problem.

Maximize $\qquad\qquad 40X + 50Y$

subject to the constraints

$$25X + 20Y \leqslant 5,000$$
$$20X + 30Y \leqslant 7,500 \qquad\qquad (3.9)$$

and

$$X, \quad Y \geqslant 0$$

The constraint set (3.9) has been graphed in Figure 3.11.

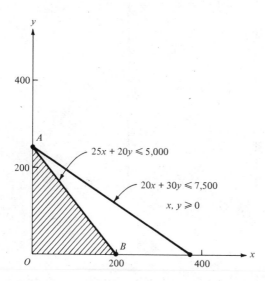

Figure 3.11

* In linear programming problems, if we can identify two optimal corner points (such as M
and H), then *any* linear combination of these points is also an optimal solution. See Section 5.6.6
for the definition of a linear combination.

The optimal solution, as the reader can verify, is at point A, which means that $X = 0$, $Y = 250$. That is, the number of positive variables in the optimal solution is 1, while the number of structural constraints, m, is 2. Hence it is a degenerate basic feasible solution. Note also that the second constraint is active just at one point in Figure 3.11. The problem of degeneracy is discussed further in Chapter 7.

3.5.3 The Unbounded Problem

The constraint set of a linear programming problem can be either bounded or unbounded. Figure 3.6 is an example of a bounded problem because the convex set $OMHL$ is closed from all sides. In such problems, there is always an *upper* (or *lower*) *bound* on the objective function. For example, the upper bound on the profit in our problem was $7,700. Most real-life linear programming problems are bounded problems. However, we can construct linear programming problems in which objective functions are not bounded.

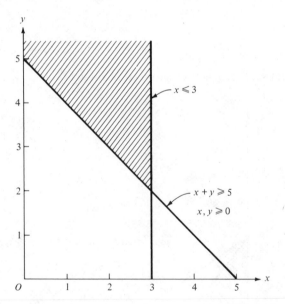

Figure 3.12

A problem is unbounded if its constraints are such that they result in a convex set that is open. Figure 3.12 shows an unbounded problem. The unbounded problem is of no practical significance. An unbounded linear programming problem can result from an improper formulation of the real-life problem.

The unbounded problem graphed in Figure 3.12 is mathematically stated as follows:

Maximize $\qquad\qquad 3X + 2Y$

subject to the constraints

$$X \qquad\ \leqslant 3$$
$$X +\ Y \geqslant 5$$

and

$$X, \qquad Y \geqslant 0$$

3.5.4 No Feasible Solution

If the linear programming problem is such that the intersection of its constraints yields an empty set, we have a case of *no feasible solution*. This happens whenever the constraints are inconsistent (i.e., mutually exclusive). Let us illustrate by the following problem.

Maximize $\qquad\qquad 10X + 20Y$

subject to the constraints

$$10X +\ 6Y \leqslant 2{,}500$$
$$10X +\ 6Y \geqslant 3{,}000$$

and

$$X, \qquad Y \geqslant \qquad 0$$

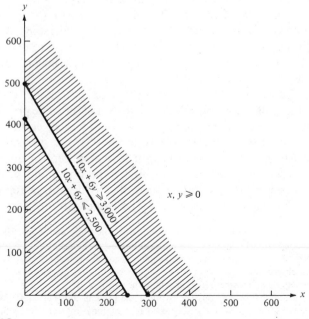

Figure 3.13

The constraints of the above problem are graphed in Figure 3.13. It is clear that no point in Figure 3.13 can simultaneously satisfy all the inequalities. Hence we have a special case in which no feasible solution can be found. The no-feasible-solution case can reflect a managerial situation in which the demands on the system cannot simultaneously be satisfied. This type of situation calls for managerial ability to resolve the impossible conflicting demands to a level that will permit organizational decisions to be made.

3.6 THE RESOURCE (OR REQUIREMENT) SPACE

Let us consider again the problem shown in Figure 3.6. Since the optimal solution, located at point H, left us with a spare capacity of 100 units of the packaging capacity, it stands to reason that, under the assumptions* of our model, the *marginal* worth of 1 unit of the packaging capacity is zero. By the same token, since the cutting and the folding capacities are fully utilized in the optimal solution, their marginal worths will be positive. This is because any additional units of either the cutting or the folding capacity will permit us to increase our production and hence the profit.

The question for the manager, therefore, is to determine the marginal worths of the resources with which he is working. This knowledge will permit him to purchase additional resources, depending upon the price levels at which they are offered. Now, one of the characteristics of linear programming is that the manager can determine these marginal worths (or shadow prices) from the optimal solution itself.† The best way to understand the meaning of the shadow prices (or marginal worths) of resources is to understand the structure of what is called duality or the dual problem in linear programming. We formulate and describe the dual in Chapter 8.

In this section we utilize the resource space to accomplish two tasks. First, by comparing the graphical representation of the same problem in two different spaces‡ we can better appreciate the ideas of constraints, fixed technology, and the set of feasible solutions. Second, utilization of the resource space provides an introduction to the concept of dual variables which we discuss in Chapter 8. Any linear programming problem that involves two constraints and any number of activities can be represented in the resource (or requirement) space.** The two constraints form the axes,

---

* Linear programming assumes certainty, linearity, and static conditions. The assumption of static conditions is most relevant in the discussion of Section 3.6.

† The simplex method, presented in Chapter 7, enables us to retrieve this information from the optimal tableau.

‡ Activity space and the resource (or requirement) space.

** For an example involving three constraints, see Loomba and Turban [1974, p. 70]. The graphical representation is obviously not useful for anything involving more than three dimensions.

and the various activities are graphed as "rays" originating from the origin. The region of feasible solution is determined by the resource constraints and the configuration of the activity "rays."

Let us consider a simple linear programming problem that involves two activities and two constraints. The technical specifications of the problem are

Table 3.4

	Product		
Resource	Carrots	Peas	Capacity
Labor	5	4	20,000 (man-hours)
Land	3	6	25,000 (square feet)
Profit per unit	$1.5	$2	

shown in Table 3.4. The problem of Table 3.4 can be expressed mathematically as follows:

Maximize $\quad\quad\quad\quad 1.5X + 2Y$

subject to the constraints

$$5X + 4Y \leqslant 20{,}000$$
$$3X + 6Y \leqslant 25{,}000 \quad\quad\quad (3.10)$$
$$X, \quad Y \geqslant \quad 0$$

Before discussing the graphing of this problem in the resource space, we graph it in the activity space as shown in Figure 3.14. Note that $OPBC$ represents the region of feasible solution. It can be shown that the optimal solution to this problem is at point $B(X = 1{,}111.1, Y = 3{,}611.1,$ and profit $= \$8{,}888.89)$.

Let us now graph the same problem in the resource space. As shown in Table 3.4, the two resources are labor (measured in man-hours) and land (measured in square feet).* These two resources form the two axes, with their origin at point O, of the resource space, as shown in Figure 3.15. First, we graph the two capacity constraints, lines DB and AB. We then plot the two product rays from the information contained in Table 3.4. Note that 1 unit of carrots requires 5 man-hours and 3 square feet of land. Thus 2,000 units of carrots require 10,000 man-hours and 6,000 square feet of land, and this is point N in Figure 3.15. A line joining points

---

* This example indicates that the general linear programming problem can deal with constraints stated in heterogeneous units.

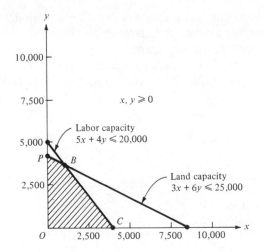

Figure 3.14

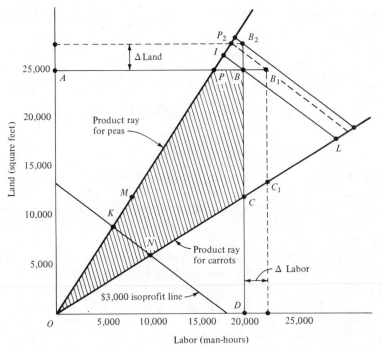

Figure 3.15

O and N defines the product ray for carrots. Note that the slope of this product ray, as well as any other product ray, is determined by technology. Also, the product rays are linear because of the assumption of fixed technology.

The second ray for peas can also be graphed from the information available in Table 3.4. For example, the point M in Figure 3.15 represents a production of 2,000 units of peas, which requires the expenditure of 8,000 man-hours and 12,000 square feet of land. By joining points O and M, we obtain the product ray for carrots. The convex set $OPBC$, shown in Figure 3.15, is the region of feasible solutions.

We can now graph an isoprofit line. For example, the point $N(X = 2,000$; i.e., labor $= 10,000$ man-hours and land $= 6,000$ square feet) corresponds to a profit of $3,000. Similarly, the point $K(Y = 1,500$; i.e., labor $= 6,000$ man-hours and land $= 9,000$ square feet) represents a profit level of $3,000. If we join points N and K, we obtain the $3,000 isoprofit line NK. Now, successively higher isoprofit lines can be constructed until the isoprofit line that is farthest from the origin, but passing through one of the extreme corner points of the convex set $OPBC$ is determined. The optimal solution, thus, will be found at one of the points P, B, or C.* In our case, the point B corresponds to the optimal solution. The isoprofit line IBL, which passes through point B, corresponds to a profit of $8,888.89.

The point B corresponds to a production program of $X = 1,111.1$ and $Y = 3,611.1$. As is clear from Figure 3.15, both the resources are fully utilized at point B, and therefore their marginal worths are positive. Specific values for the marginal worths of the two resources can now be determined in the following manner.

First, increase the availability of labor from 20,000 man-hours to 22,500 man-hours, but keep the land capacity at 25,000 square feet.† This will yield a new convex set OPB_1C_1 and a new optimal solution at point $B_1(X = 1,944\frac{4}{9}, Y = 3,194\frac{4}{9})$, as shown in Figure 3.15. The new profit level is $9,305.56. Since the $416.67 increase in profit was made solely due to an increase of 2,500 man-hours of labor, we conclude that the marginal worth of labor is $0.1667 per man-hour. Second, increase the availability of land from 25,000 square feet to 27,500 square feet but keep the labor capacity at 20,000 man-hours. This will yield a new convex set OP_2B_2C and a new optimal solution at point B_2 ($X = 555\frac{5}{9}, Y = 4,305\frac{5}{9}$). The new profit level is $9,444.44. Since the $555.55 increase in profit was made solely due to an increase of 2,500 square feet of land, we conclude that the marginal worth of labor is $0.222 per square foot.

* If the isoprofit line is parallel to any of the boundary lines of the convex set, we have more than one optimal solution.

† This simple method requires that the new optimal solution will remain such that both resources are exhausted. In other words, the increase in one resource should not be of such a magnitude that it will shift the optimal solution to extensions of point P or C, where only one resource is fully utilized.

Note that if the optimal solution to our original problem had been at point *P*, the marginal worth of labor would have been zero. Similarly, had the optimal solution been at point *C*, the marginal worth of land would have been zero.

Since the marginal worths are not constants but instead are determined by the relationship of the optimal solution to the available resource capacities, they are variable. The variables that represent the values of the marginal worths are known as *dual variables* (discussed in Chapter 8). Note the correspondence of the convex set *OPBC*, as shown in Figures 3.14 and 3.15.

3.7 PROCEDURE SUMMARY FOR THE GRAPHICAL METHOD (A TWO-DIMENSIONAL MAXIMIZATION PROBLEM)

STEP 1

Express mathematically the given specifications of the problem into three parts: (1) a linear objective function, (2) a set of structural constraints (consisting of inequalities or equalities), and (3) a set of nonnegativity constraints.

STEP 2

Graph each structural constraint and each nonnegativity constraint in the activity space. The common intersection of these constraints will yield a convex polygon that contains all feasible solutions.

STEP 3

By choosing a convenient profit figure, draw an isoprofit line so that it falls within the region of feasible solutions (shaded area).

STEP 4

Move this isoprofit line parallel to itself and farther from the origin,* until an optimal solution is determined.

---

* We assume, of course, that each of the two coefficients in the objective (profit) function is positive. Let us consider two other cases: (1) Both of the coefficients in the objective function are negative (e.g., maximize $-10X - 15Y$), and (2) one of the coefficients in the objective function is negative while the other is positive (e.g., maximize $-10X + 15Y$). If, in a maximization problem, both the coefficients in the objective function are negative (as in case 1), it is obvious that the level of each activity should be reduced to zero. In other words, neither X nor Y should be produced. If one of the coefficients in the objective function is positive while the other is negative (as in case 2), it is again obvious that the product with the negative coefficient (product X) should not be produced at all, whereas as much as possible of the product with the positive coefficient (product Y) should be produced within the given constraints.

The graphical method can also be used for solving a minimization problem. Of the four steps listed, only the last two need be revised to obtain a complete procedure for solving a two-dimensional minimization problem. In step 3, a convenient cost figure, rather than a profit figure, is chosen, and a corresponding isocost line is drawn. In step 4, the isocost line is moved parallel to itself but closer to the origin, until an optimal solution is determined.

3.8 SUMMARY

In this chapter we employed a graphical approach in solving a linear programming problem. The same problem was graphed in the activity space as well as the resource space. The activity space representation permitted us to define different types of solutions, such as feasible, basic feasible, and degenerate basic feasible solutions. The set of solutions was found to be convex, and it was indicated that the number of solutions is infinite. The resource space representation was helpful in clarifying the concept of "marginal worth" of a resource.

The usual case in linear programming problems is that the optimal solution can be found and is unique. There are, however, certain special cases, such as (1) multiple optimal solutions, (2) the degenerate basic feasible solution, (3) the unbounded solution, and (4) no feasible solution. Each of these special cases was illustrated graphically.

The graphical approach is of limited value and is not an efficient method of solving linear programming problems. However, as we discovered in this chapter, it is very useful in clarifying several important concepts in linear programming. The algebraic approach is more general and is discussed in the next chapter.

REFERENCES

Adams, W. J., et al. *Elements of Linear Programming*. New York: Van Nostrand Reinhold, 1969.

Cooper, L., and D. Steinberg. *Linear Programming*. Philadelphia: W. B. Saunders Company, 1974.

Hughes, A. J., and D. Grawiog. *Linear Programming*. Reading, Mass.: Addison-Wesley Publishing Co., Inc., 1973.

Loomba, N. P. *Linear Programming*. New York: McGraw-Hill Book Company, 1964.

Loomba, N. P., and E. Turban. *Applied Programming for Management*. New York: Holt, Rinehart and Winston, Inc., 1974.

McMillan, C. *Mathematical Programming*. New York: John Wiley & Sons, Inc., 1974.

Strum, J. E. *Introduction to Linear Programming*. San Francisco: Holden-Day, Inc., 1972.

REVIEW QUESTIONS AND PROBLEMS

3.1. Describe the advantages and the limitations of the graphical method in solving linear programming problems.

3.2. Graph the following inequalities:

$$1X + 1Y \geqslant 4$$
$$6X - 12Y \geqslant 12$$
$$5X + 8Y \leqslant 40$$
$$X \geqslant 0, \qquad Y \geqslant 0$$

(a) Identify the polygon enclosed by these inequalities.

(b) Draw a straight line joining *any* two points in the polygon. Does this line lie completely within the polygon? What is the significance of your answer?

3.3. The management of National Oil Company has received orders from International Airlines to supply 500 units of gasoline *A* and 750 units of gasoline *B*. National Oil Company has two processes that can be used (processes I and II) for the manufacture of gasolines *A* and *B*. One hour of process I produces 5 units of gasoline *A* and 10 units of gasoline *B*. One hour of process II produces 8 units of gasoline *A* and 6 units of gasoline *B*. Process I uses a particular combination of crudes which cost $7 per hour, and process II uses a different combination of crudes, which costs $9.20 per hour. Find, graphically, the optimal mix of process I and process II.

3.4. Richards Lumber Company produces three major products; shelves, doors, and wooden beams for construction. The profit margin, per dozen, is as follows: shelves, $60; doors, $50; and beams, $50. All products must go through two processes, cutting and sanding, which have an upper capacity of 40 and 50 hours per week, respectively. The three products require process time, as shown in Table 3.5. (Time is expressed in hours per dozen units.) How can the company maximize its profits? Use the resource-space concept.

Table 3.5

	Shelves	Doors	Beams	Weekly Capacity
Cutting	2	$1\frac{1}{4}$	1	40
Sanding	1	1	$1\frac{2}{3}$	50

3.5. Solve the following linear programming problem graphically:

Minimize $\qquad\qquad\qquad 1X + 1Y$

subject to the constraints

$$3.5X + Y \leqslant 7$$
$$-0.5X + Y \geqslant 1$$
$$-8X + 10Y \leqslant 40$$
$$X \qquad Y \geqslant 0$$

Will the solution change if the right-hand side of first inequality is changed from 7 to 5?

3.6. Maximize $\qquad\qquad\qquad 4Y + 5X$

subject to the constraints

$$3X - 5Y \geqslant 15$$
$$14X + 6Y \geqslant 42$$
$$X, \quad Y \geqslant 0$$

(a) Show graphically that the above problem is unbounded.
(b) What would be the solution if the objective function were to be minimized instead of being maximized?

3.7. The Columbian Coffee Company produces two different kinds of coffee using two different types of beans (call them A and B). Mixture I must have exactly 10 ounces of A and 6 ounces of B. Mixture II must have 8 ounces of A and 8 ounces of B. The company has 60,000 ounces of A and 40,000 ounces of B at its disposal. Mixtures I and II return profits of $0.75 and $1.00, respectively. Determine the optimal combination of I and II.

3.8. A manufacturing firm sells two types of plastic containers for household consumption. The firm's market potential for these containers is unlimited. Each container must be processed through two different machines. The relevant data on machine capacities, processing times, and profit contribution per unit are given in Table 3.6. Use the graphical method to determine the optimal product mix under the following conditions:

(a) For the data given in Table 3.6.
(b) The profit contribution of container A becomes 15 cents per unit.
(c) Containers A and B yield negative profit contributions (losses) of 10 cents and 5 cents, respectively.
(d) The capacity of machine 1 is increased to 2,500 minutes.

Table 3.6

Machine	Processing Time (minutes) per Unit of:		Available Capacity per Time Period (minutes)
	Container A	Container B	
1	4	2	2,000
2	3	5	3,000
Profit contribution per unit (cents)	20	10	

3.9. There are two types of vitamin pills available on the market, types I and II. Pill I contains 0.12 mg of vitamin A and 0.6 mg of vitamin B. Pill II contains 0.14 mg of vitamin A and 0.16 mg of vitamin B. Costs of pills I and II are 29 cents per pill and 23 cents per pill, respectively. Minimum daily adult requirements, as set by the Food and Drug Administration, are 1 milligram of vitamin A and 10 milligrams of vitamin B. Determine the optimum mixture of pills in order to meet half the daily adult requirement. Solve the problem if full daily adult requirements were to be met by the pills.

4

Linear Programming:
An Algebraic Approach

4.1 INTRODUCTION

The graphical method of solving the linear programming problem is not a powerful method. Its utility is essentially limited to problems in which the structural constraint set has a dimension no larger than $2 \times n$ or $m \times 2$. Furthermore, the graphical method is cumbersome and is certainly not an efficient method. The algebraic approach is more general and, theoretically speaking, can handle problems of any size. In this chapter we shall describe the algebraic approach by solving the same problem that we used in illustrating the graphical method. The problem is outlined in Table 4.1. This problem, as the reader will recall, was graphed in the activity space, as shown in Figure 3.6.

4.2 TRANSFORMATION OF INEQUALITIES INTO EQUALITIES

The transformation of inequalities into equalities is the first step in the algebraic approach. We explain in this section how these transformations are made. There are essentially two types of inequalities: (1) less than or equal to ($\leqslant$), and (2) greater than or equal to ($\geqslant$). The "less than or equal to" inequality is used to indicate the upper limit of a resource. For example, the

Table 4.1

Department	Product		Capacity per Time Period
	A	B	
Cutting	10	6	2,500
Folding	5	10	2,000
Packaging	1	2	500
Profit per unit	$23	$32	

given capacity levels of the cutting, folding, and packaging departments in Table 4.1 are the upper limits of the three resources. The "greater than or equal to" inequality is used to indicate the lower limit of a characteristic or requirement. For example, in the diet problem of Section 1.9.2, the requirement levels for nutrients I and II are the lower limits.

The specifications of our product-mix problem are listed in Table 4.1. The resource limitation for the cutting department can be expressed mathematically as

$$10X + 6Y \leqslant 2,500 \qquad (4.1)$$

It is obvious that the values of X and Y must be so chosen that when inserted in (4.1), the left-hand side of the inequality must be *less than* or *equal to* the right-hand side. If a set of values (say $X = 100$, $Y = 100$) is such that the left-hand side is indeed less than the right-hand side, we observe that the cutting resource is not fully utilized and that some *slack* is left. For example, for $X = 100$, $Y = 100$, the slack (or the idle capacity) is 900 units. The magnitude of the slack is a variable because its value is determined by what values are assigned to X and Y. For example, if $X = 100$, $Y = 50$, the slack is 1,200 units. Note that the slack cannot be negative.

For any set of values X and Y, the *addition* of the slack variable to the left-hand side will make the left-hand side equal to the right-hand side.* Hence the rule:

An inequality of the "less than or equal to" type is transformed into an equality by the addition of a nonnegative slack variable.

We shall denote slack variables with the letters $S_1, S_2, S_3, \ldots$. The slack variables can be thought of as imaginary products, each requiring for its production 1 unit of capacity from *only one* of the resources and 0 units of capacity from the others, and each yielding a profit of zero. The production of 1 unit of S_1, for example, requires 1 unit of cutting capacity but 0 units of folding capacity and 0 units of packaging capacity.

---

* If $X = 100$ and $Y = 100$, then the slack is 900. Hence, according to (4.1), $10(100) + 6(100) + 900 = 2,500$.

Let us now consider the diet problem of Section 1.9.2. The requirement for nutrient I can be expressed mathematically as

$$2X + 3Y \geqslant 3{,}500 \tag{4.2}$$

Again, it is obvious that the values of X and Y must be so chosen that when inserted in (4.2) the left-hand side of the inequality must be greater than or equal to the right-hand side. If the X and Y values are such that they make the left-hand side greater than the right-hand side, the equality can be restored by *subtracting* a nonnegative *surplus* variable. A specific value of the surplus variable represents the excess of nutrient I over the required level of 3,500. This argument leads us to the following rule:

An inequality of the "greater than or equal to" type is transformed into an equality by the subtraction of a nonnegative surplus variable.

We shall denote surplus variables by the letters $R_1, R_2, R_3, \ldots$.

4.3 RATIONALE FOR THE ALGEBRAIC APPROACH

Let us think of a linear programming problem with m structural constraints and n activities (real variables). If the m constraints are stated as inequalities, we can convert them into equalities by the addition or subtraction of a total of m slack or surplus variables. Then we have a system of m equations with $m + n$ variables.* Since the number of equations is less than the number of variables (m is less than $m + n$), we have an infinite† number of solutions. This property, that a linear programming problem has an infinite number of solutions, was also demonstrated in the graphical method (see Section 3.3.3). Furthermore, it was stated in Section 3.3.3 that in order to locate the optimal solution, we have to search only the extreme (or corner) points of the convex set. This property of only searching the corner points reduces the search process to a finite effort, but even then there is a very large number of such corner points. For example, the number of possible corner points in an $m \times (m + n)$ system is given by the combination of "$(m + n)$ things taken m at a time." ‡

* The $m + n$ variables consist of n real variables and m slack or surplus variables.

† For example, consider the single equation system $X + Y = 10$. In this system, the number of equations (one) is less than the number of variables (two); and hence we have an infinite number of solutions. The reader can check that for every arbitrary value of X, there is a value of Y so that $X + Y = 10$.

‡ This is calculated as

$$\binom{m + n}{m} = \frac{(m + n)!}{(m!)(m + n - m)!}$$

For our example of Table 4.1, then, the number of possible corner points will be "combination of 5 things taken 3 at a time." That is, $5!/(3!2!) = 10$. Of these, only four corner points, O, M, H, and L, represent feasible solutions.

Furthermore, not all the corner points are feasible solutions. For example, in Figure 3.6 only the four corner points, O, M, H, and L, represent feasible solutions. The algebraic method is so designed that it tests only the feasible solution at the corner points of the convex set.* Actually, two things are accomplished. First, a basic feasible solution (or program) is identified or designed, and tested for optimality. Second, if a specific program is not optimal, it is revised to yield a better solution, which is again tested for optimality. This iterative procedure is repeated until the optimal solution is identified.

In the algebraic method, as will be illustrated in the next section, the objective function is used to test the optimality of a given solution. The objective function is modified to yield information as to (1) whether or not the given program can be improved and (2) how to design a new program. Thus the difficult or even impossible task of determining all possible production combinations need not be undertaken. Instead, we can design an initial program such that the given constraints are not violated. The initial program can then be tested for optimality by examining the associated objective function.† If the test indicates that a better program can be designed, the initial program is revised. The revised program is again tested for optimality, and if further improvement in the objective function is possible, another program is designed. This process is repeated until an optimal solution has been obtained.

4.4 THE ALGEBRAIC APPROACH—AN ILLUSTRATIVE EXAMPLE

We provide substance to the ideas of Section 4.3 by solving the problem of Table 4.1 by the algebraic method. Our task, as before, is to find the optimal mix that will yield maximum profit.

4.4.1 Express the Given Data as a Linear Programming Problem

Our first task is to express the given data as a linear programming problem consisting of a linear objective function, a set of structural con-

---

*It will be recalled from Section 3.3.3 that we termed a corner-point solution as either a basic feasible or a degenerate basic feasible, depending upon the number of positive variables in the solution in relation to the number of structural constraints m.

†For the maximization case, optimality requires that the coefficients of each variable included in the associated objective function be negative. For the minimization case, optimality requires that the coefficient of each variable included in the associated objective function be positive.

straints, and a set of nonnegativity constraints. The objective function for the problem is

Maximize $23X + 32Y$

All the structural constraints are inequalities of the "less than or equal to" type. These inequalities are

for the cutting department

$$10X + 6Y \leqslant 2{,}500 \tag{4.3}$$

for the folding department

$$5X + 10Y \leqslant 2{,}000 \tag{4.4}$$

for the packaging department

$$1X + 2Y \leqslant 500 \tag{4.5}$$

The nonnegativity constraints are

$$X \geqslant 0 \qquad Y \geqslant 0$$

4.4.2 Transform the Inequalities into Equalities and Modify the Objective Function

Since all three inequalities (4.3) to (4.5) are of the "less than or equal to" type, they can be transformed into equalities by the addition of nonnegative slack variables S_1, S_2, and S_3. Remember that each unit of any slack variable contributes a zero profit. Hence the slack variables can be inserted in the objective function provided that we assign a value of zero to their profit coefficients. Thus our problem can be stated as follows:

Maximize $23X + 32Y + 0S_1 + 0S_2 + 0S_3$

subject to the constraints

$$10X + 6Y + S_1 \qquad\qquad = 2{,}500 \tag{4.6}$$
$$5X + 10Y \qquad + S_2 \qquad = 2{,}000 \tag{4.7}$$
$$1X + 2Y \qquad\qquad + S_3 = 500 \tag{4.8}$$

and

$$X, \quad Y, \quad S_1, \quad S_2, \quad S_3 \geqslant 0$$

The interpretation of the slack variables is the same as given in Section 4.2.

4.4.3 Design an Initial Program

Let us develop a program in which we propose to produce only the "imaginary" products S_1, S_2, and S_3.* This means that in Equations (4.6) to (4.8) we let $X = 0$ and $Y = 0$. Our initial program, therefore, is to produce

* This will correspond to starting from point O in Figure 3.6.

2,500 units of S_1, 2,000 units of S_2, and 500 units of S_3. Thus the initial program consists of the following values of the different variables:

$$X = 0 \qquad Y = 0 \qquad S_1 = 2{,}500 \qquad S_2 = 2{,}000 \qquad S_3 = 500$$

The initial program, together with other information, is contained in the following equations, which are obtained by a simple rearrangement of Equations (4.6) to (4.8): *

Program 1

$$
\begin{aligned}
S_1 &= 2{,}500 - 10X - 6Y & (4.9) \\
S_2 &= 2{,}000 - 5X - 10Y & (4.10) \\
S_3 &= 500 - 1X - 2Y & (4.11)
\end{aligned}
$$

Note that Equations (4.9) to (4.11) imply the following production program:

$$S_1 = 2{,}500 \qquad S_2 = 2{,}000 \qquad S_3 = 500$$

and

$$X = 0 \qquad Y = 0$$

These equations have physical interpretations. For example, Equation (4.9) says that if $X = 0$ and $Y = 0$, we shall produce 2,500 units of S_1. That is, all the cutting-department capacity will remain idle. Further, Equation (4.9) reveals that if we want to produce, at this stage, say, 1 unit of X, we must be willing to sacrifice 10 units of S_1. Similarly, introduction of 1 unit of Y will demand a reduction of S_1 by 6 units. In other words, Equation (4.9) gives us information regarding the physical ratios of exchange or substitution between X and S_1, and Y and S_1. These physical ratios of substitution at this stage are nothing but the coefficients of the variables X and Y in the preceding set of linear equations. Similar comments can be made with respect to Equations (4.10) and (4.11).

We should also keep in mind, as Equations (4.9) to (4.11) reveal, that the introduction of 1 unit of, say, X requires not only the reduction of S_1 by 10 units but the simultaneous reduction of S_2 by 5 units and S_3 by 1 unit. This is, of course, determined by the manufacturing requirements of the products, as given in Table 4.1.

We note that the addition of a unit of X or Y makes simultaneous demands for the reduction of S_1, S_2, and S_3. Therefore, the total amount of

---

*Each equation is solved for the product *included* in a given program. Thus, in this case, S_1, S_2, and S_3 are brought to the left-hand sides of the equations. The variables on the left-hand sides of the program are the program or *basic* variables. The value of each basic variable is given by the constant term on the right-hand side. The variables on the right-hand sides are *nonbasic* variables. Their values are assumed to be zero. Their coefficients represent ratios of substitution between them and the basic variables.

X or Y that can, at this stage, be *brought in* the program is limited by the current magnitudes of S_1, S_2, and S_3.

Is our initial program an optimal program? The answer is obviously no, because all the resources are idle and there is no *real* production. However, later we shall actually conduct the optimality test by examining the signs of the profit coefficients of the *modified objective function* that is obtained by substituting one of the program equations [in this case the program equations are (4.9) to (4.11)] in the preceding objective function (at this stage, $23X + 32Y + 0S_1 + 0S_2 + 0S_3$). Since the coefficients of S_1, S_2, and S_3 are zero, the modified objective function is obviously:

$$\text{profit} = \boxed{23X + 32Y} \qquad \text{objective function associated with program 1}$$

Since both the profit coefficients in the modified objective function shown above are positive, program 1 is not optimal.

4.4.4 Revise the Initial Program and Test for Optimality

Insofar as the initial program gives a profit of zero, it can certainly be improved. This improvement is made by designing a new program in such a way that *at least one* of the variables (products) in the present program is replaced by *exactly one* of the variables (products) not in the present program. The replacement (or exchange) is carried out by first identifying the *incoming* variable and then identifying the *outgoing* variable(s).

In our example, the variables (products) included* in the first program are S_1, S_2, and S_3; the variables (products) external† to the first program are X and Y. Thus revision of the first program means that one of the variables (products) S_1, S_2, and S_3 must be replaced by either X or Y. Two questions, in other words, must be answered:

1. Which one of the variables (products) not in the present program should be brought in to replace one of the variables (products) currently in the program? That is, we identify the *incoming variable*.

2. What is the maximum amount of the chosen incoming variable (product) that can be brought in? The answer to this question will also identify the *outgoing variable* (or the product to be replaced).

To identify the particular product to be brought in, we must make a comparison of the cost or profit consequences associated with the introduction of 1 unit of each such product that is currently not included in the program. Then the variable (product) with the highest net advantage per

---

* Variables that are included in the program are called *basic variables*. In the initial program, therefore, S_1, S_2, and S_3 are basic variables.

† Variables that are not in the program are called *nonbasic variables*. In the initial program X and Y are nonbasic variables.

unit is chosen to be included in the revised program. For example, examination of our initial program [contained in Equations (4.9) to (4.11)] shows that producing 1 unit of Y will require sacrificing 6 units of S_1, 10 units of S_2, and 2 units of S_3. Insofar as S_1, S_2, and S_3 have profit contributions of zero per unit while Y has a profit contribution of 32 per unit, the introduction of Y at this stage would be a desirable course of action. Similar comments can be made in connection with introducing 1 unit of X.

While revising a given program, the important question to be asked is: What do we gain by bringing in, say, 1 unit of a particular variable (product) not in the current program (i.e., nonbasic variable), and what do we lose by sacrificing some corresponding* quantities of the variables (products) included in the current program (i.e., basic variable)? If the gain is more than the loss, the substitution is desirable; otherwise, it is not. At this stage of our solution, the net advantage associated with the introduction of 1 unit of Y is 32, while 1 unit of X will yield a net advantage of 23. We therefore would want to introduce the variable (product) Y in our next program. That is, Y has been identified as the incoming variable.

Another way to reach the same conclusion is to incorporate the equations representing the present program in the corresponding objective function. In this manner, we can immediately identify those variables which can be used for revising the current program to give a net increase in profit. Since this is a maximization problem, these variables will be those which have positive coefficients in the modified objective function. Of these, the variable with the highest net advantage (largest positive coefficient) is chosen as the incoming variable to be introduced in the next program.

This approach for selecting the incoming variable by utilizing the modified objective function associated with a given program is mechanical in nature and quite easy to use.

The modified objective function associated with program 1 has already been derived:

$$\text{profit} = \boxed{23X + 32Y} \qquad \text{objective function associated with program 1}$$

Since, in this function, Y has the largest positive coefficient, it is the variable (product) to be first introduced into the solution. That is, we have identified Y as the incoming variable. At this stage of the solution, therefore, Y is the key variable.† This answers our first question: Which product is to be brought in?

To answer the second question, concerning the maximum amount of the chosen incoming product, we reason as follows. It has been established that the net advantage associated with the introduction of 1 unit of Y in our next program, at this stage, is 32. Since this is a profitable course of action,

---

* The corresponding quantities are determined by the ratios of substitution operating at a given solution stage and contained in equations that represent a given program.

† See Chapter 7 for the corresponding concept of key column.

we should continue to bring in Y until one of the currently produced products S_1, S_2, and S_3 (idle capacities) is eliminated from the initial program. The production of Y beyond the level determined in this manner is, of course, not possible, for that would violate some of the nonnegativity constraints. That is, the given resource capacities could not support the production of Y beyond the level determined in the above-mentioned manner.

Therefore, to determine the maximum possible level of Y that can be produced in order to improve the initial program, let us examine Equations (4.9) to (4.11). These equations, as stated earlier, represent our initial program involving the production of $S_1 = 2,500$ units, $S_2 = 2,000$ units, and $S_3 = 500$ units. Knowing that only Y is to be brought in the solution, we set $X = 0$ in Equations (4.9) to (4.11) and increase the value of Y until the left-hand side of one of these equations becomes zero. That is, the amount of Y to be brought in is that which will make some S_i (S_1 or S_2 or S_3) the first to become zero. This would mean that the chosen "incoming" variable (product) Y has completely eliminated one of the variables (S_1, S_2, or S_3) in the current program. In order to determine which of the three variables (S_1, S_2, or S_3) will be eliminated from program 1 (i.e., the "outgoing" variable), we set $X = 0$, $S_1 = 0$, $S_2 = 0$, and $S_3 = 0$ in Equations (4.9) to (4.11). We find that the variable (product) S_2 will be eliminated from the current program and that the maximum amount of Y that can be introduced into the next program is 200 units: From Equation (4.9),

$$\text{maximum } Y = \tfrac{2,500}{6} = 416\tfrac{2}{3} \text{ units}$$

From Equation (4.10),

$$\text{maximum } Y = \tfrac{2,000}{10} = 200 \text{ units } \checkmark$$

From Equation (4.11),

$$\text{maximum } Y = \tfrac{500}{2} = 250 \text{ units}$$

Thus we have answered our second question, namely, "What is the maximum amount of the chosen 'incoming' product that can be brought in?" As indicated by the check mark, $\checkmark$, the maximum amount of the incoming product, Y, in view of the nonnegativity constraints, is 200 units.

The above calculations also indicate that the variable (product) S_2 will be replaced by Y in the next program. Equation (4.10), therefore, is what is called the limiting or key equation.*

Since Y is going to replace S_2, we rearrange Equation (4.10) so that the variable Y is on the left-hand side. Thus

$$10Y = 2,000 - 5X - S_2$$

or

$$Y = 200 - \tfrac{1}{2}X - \tfrac{1}{10}S_2 \tag{4.12}$$

---

* See Chapter 7 for the corresponding concept of key row.

Equation (4.12) can be interpreted as follows. Assuming that $X = 0$, and $S_2 = 0$, we shall be producing 200 units of Y. Furthermore, this equation gives information, as previously explained, on the ratios of substitution (at this solution stage) between X, S_2 and Y.

The production of Y, in addition to reducing the production of S_2 to zero, also reduces the amounts of S_1 and S_3 produced. We can get this information in precise terms by substituting Equation (4.12) in Equations (4.9) and (4.11).

Substituting (4.12) in Equation (4.9), we obtain

$$S_1 = 2,500 - 10X - 6\left(200 - \tfrac{1}{2}X - \tfrac{1}{10}S_2\right)$$
$$= 1,300 - 7X + \tfrac{3}{5}S_2 \tag{4.13}$$

and substituting (4.12) in Equation (4.11),

$$S_3 = 500 - X - 2\left(200 - \tfrac{1}{2}X - \tfrac{1}{10}S_2\right)$$

or

$$S_3 = 100 + 0X + \tfrac{1}{5}S_2 \tag{4.14}$$

Equation (4.13) gives us the number of S_1 units remaining (idle capacity of cutting department) as well as the pertinent ratios of substitution among the different variables at this stage. Equation (4.14) gives us similar information in connection with the packaging department. Our revised program (program 2), therefore, is represented by the following equations:

Program 2

$S_1 = 1,300 - 7X + \tfrac{3}{5}S_2$	(4.13)
$Y = \quad 200 - \tfrac{1}{2}X - \tfrac{1}{10}S_2$	(4.12)
$S_3 = \quad 100 + 0X + \tfrac{1}{5}S_2$	(4.14)

Equations (4.12) to (4.14) imply the following:

$$Y = 200 \qquad S_1 = 1,300 \qquad S_3 = 100$$

and

$$X = 0 \qquad S_2 = 0$$

Is our second program an optimal program? As explained earlier, the optimality of a given program is tested by examining the signs of the profit coefficients in the modified objective function. We obtain the modified objective function by substituting in the objective function associated with the preceding program the equation that relates the incoming variable to the nonbasic variables of the current program. For example, while revising program 1, we observed that Equation (4.10) was the limiting or key equation (i.e., Y was identified as the incoming variable). This was rearranged to yield

Equation (4.12). Thus we test the optimality of the current program (program 2) by substituting (4.12) into the objective function associated with program 1:

$$\text{profit} = 23X + 32Y$$

$$= 23X + 32(200 - \tfrac{1}{2}X - \tfrac{1}{10}S_2)$$

$$= \boxed{6,400 + 7X - \tfrac{16}{5}S_2} \qquad \begin{array}{l}\text{objective function associated} \\ \text{with program 2}\end{array} \qquad (4.15)$$

This modified profit function gives us the following information: program 2 gives a total profit contribution of $6,400. An additional unit of X, if it can be produced at this stage within the given constraints, will bring a *net* advantage of $7; but the addition of 1 unit of S_2 at this stage will subtract $3.2 from the profit function.

It is important to know why, in Equation (4.15), the profit contribution of 1 unit of X has become $7, as compared with $23 as given in Table 4.1. The reason for this becomes clear if we keep in mind that Equation (4.15) represents the modified objective function corresponding to the solution stage represented by program 2, which has fully utilized the folding capacity. Thus the production of 1 unit of X at this stage will mean that 5 units of folding capacity (required for producing 1 unit of X) must be reallocated from Y. Since each unit of Y requires 10 units of folding capacity, Y's current level of production will thus be reduced by $\tfrac{5}{10}$ unit, which will mean a reduction of $16 ($\tfrac{5}{10} \times \$32 = \$16$) in the profit contribution. Thus the introduction of 1 unit of X at this stage results in a gross increase in profit of $23 (attributable to X) but causes a decrease of $16 in the profit contribution of Y. Hence the introduction of one unit of X at this stage will give a net profit of $7, which is the coefficient of the variable X in Equation (4.15). Similar interpretations can be given to the coefficients of the variables in the objective function at different solution stages.

Insofar as the objective function associated with a given program has variables with positive coefficients, that program can be improved. As this is the case in Equation (4.15), program 2 is not an optimal program. Note that X has the only positive coefficient in Equation (4.15), and hence we should introduce X in the next program. That is, X will now be the incoming variable.

4.4.5 Design Another Improved Program (Revision of Program)

According to the guiding rule that we established earlier (i.e., incoming variable is identified by the largest positive coefficient in the objective function associated with a given program), we now propose to introduce X into the solution. In determining the maximum amount of X that can be brought in without violating the nonnegativity constraints, we examine

Equations (4.12) to (4.14), which represent the second program. Assuming Y, S_1, S_2, and S_3 to be zero, we have, from Equation (4.12),

$$\text{maximum } X = \tfrac{200}{0.5} = 400$$

From Equation (4.13),

$$\text{maximum } X = \tfrac{1{,}300}{7} = 185\tfrac{5}{7} \checkmark$$

Equation (4.14), however, does not impose any maximum limit.* Hence Equation (4.13) provides the limiting case, and the maximum amount of X that can be brought in is $\tfrac{1{,}300}{7}$ units. The rationale for this procedure, as the reader will recall, was explained when we revised the initial program to obtain program 2. Note that while the maximum amount of the chosen incoming variable (product) that can be brought in the solution is being determined, all variables except the incoming variable (at this stage, the incoming variable is X) are given a value of zero in the set of equations that represents the current program.

Since Equation (4.13) is the limiting equation, the introduction of X must, of course, completely eliminate S_1 from the solution. Rearranging Equation (4.13),† we get

$$7X = 1{,}300 - S_1 + \tfrac{3}{5}S_2$$

or

$$X = \tfrac{1{,}300}{7} - \tfrac{1}{7}S_1 + \tfrac{3}{35}S_2 \qquad (4.16)$$

The reader, by now, must have learned to appreciate the information that can be obtained from these equations. Equation (4.16), for example, informs us that if $S_1 = 0$ and $S_2 = 0$, we shall be producing $\tfrac{1{,}300}{7}$ units of X. But what about Y and S_3? To answer this question, we substitute Equation (4.16) in (4.12) and then put Equation (4.16) in (4.14). Thus

$$Y = 200 - \tfrac{1}{2}(\tfrac{1{,}300}{7} - \tfrac{1}{7}S_1 + \tfrac{3}{35}S_2) - \tfrac{1}{10}S_2$$

or

$$Y = \tfrac{750}{7} + \tfrac{1}{14}S_1 - \tfrac{1}{7}S_2 \qquad (4.17)$$

---

* Equation (4.14) provides no limit, as the ratio of substitution between S_3 and X in Equation (4.14) is zero. If calculated, maximum X permitted, in view of Equation (4.14), will be infinity ($\tfrac{100}{0}$). Also, any negative ratio of substitution will not create any limitations. Hence the rule: When calculating the maximum amount of the incoming variable, only positive ratios of substitution need be examined. We shall utilize this rationale in the simplex method, presented in Chapter 7.

† As explained earlier, the equation to be rearranged is always that which relates the incoming variable (product) to the product to be completely eliminated from the current program. This, of course, means that the *limiting equation* is rearranged.

As the reader can verify, substituting Equation (4.16) in (4.14) does not change things, since the coefficient of X is zero in Equation (4.14). Hence program 3 is represented by the following three equations:

Program 3

$$X = \tfrac{1,300}{7} - \tfrac{1}{7}S_1 + \tfrac{3}{35}S_2$$
$$Y = \tfrac{750}{7} + \tfrac{1}{14}S_1 - \tfrac{1}{7}S_2$$
$$S_3 = 100 + 0X + \tfrac{1}{5}S_2$$

Program 3 consists of

$$Y = \tfrac{750}{7} \qquad X = \tfrac{1,300}{7} \qquad S_3 = 100$$

and

$$S_1 = 0 \qquad S_2 = 0$$

Is this the optimum program? To answer this question, we again derive and examine the modified objective function associated with program 3. Note that X is the incoming variable [and Equation (4.16) relates it to nonbasic variables in the current program] and Equation (4.15) is the profit function associated with the preceding program. Hence we substitute Equation (4.16) in (4.15):

$$\text{profit} = 6,400 + 7(\tfrac{1,300}{7} - \tfrac{1}{7}S_1 + \tfrac{3}{35}S_2) - \tfrac{16}{5}S_2$$

$$= \boxed{7,700 - S_1 - \tfrac{13}{5}S_2} \qquad \begin{array}{l}\text{objective function associated} \\ \text{with program 3}\end{array} \qquad (4.18)$$

We observe that the present program (program 3) gives a total profit of $7,700. Furthermore, an examination of the modified objective function [Equation (4.18)] indicates that coefficients of all the variables are negative. Hence program 3 is the optimal program. This is the same optimal program that was obtained in Chapter 3 by the graphical method.

Note the particular pattern of the algebraic method. Once the incoming variable (product) has been identified, it is for that variable that the limiting equation is solved. This solution is inserted into the remaining equations representing a particular program and into the preceding objective function. This pattern systematically repeats itself from program to program until the optimal solution is determined.

A summary of the various stages of the solution derived by the algebraic method is related to the corresponding corner-point solutions of Figure 3.6 in Table 4.2.

Table 4.2

Program	Variables in Solution (Basic Variables)	Variables Not in Solution (Nonbasic Variables)	Corresponding Solution from Figure 3.6	Total Profit
1	S_1, S_2, S_3	X, Y	O	$0
2	S_1, Y, S_3	X, S_2	M	6,400
3	X, Y, S_3	S_1, S_2	H	7,700

4.5 PROCEDURE SUMMARY FOR THE ALGEBRAIC APPROACH (Maximization Case)

STEP 1. Formulate the problem.

a. Make a precise statement of the objective function and translate the technical specifications of the problem into inequalities. It is assumed that all inequalities are of the "less than or equal to" type and all variables are restricted to nonnegative values.

b. Convert the inequalities into equalities by the addition of non-negative slack variables. Attach a per unit profit of zero to each slack variable or imaginary product.

STEP 2. Design an initial program.

Design an initial program so that only the *imaginary* products are being produced, that is, only the *slack* variables are included in the solution. Represent the initial program by arranging the equations of step 1 such that the products being produced (i.e., basic variables) are on the left-hand sides.

STEP 3. Revise the current program.

a. *Identify the incoming variable.* Insofar as the initial program consists of only the imaginary products (slack variables), its profit contribution is zero. Thus, to improve the initial program, the variable with the largest positive coefficient is chosen as the incoming variable. For programs other than the initial program, the incoming variable is identified by step 4b.

b. *Determine the maximum quantity of the incoming variable.* From the equations representing the current program, determine the limiting or key equation that will indicate the maximum quantity of the chosen incoming variable that can be introduced into the solution without violating the non-negativity constraints.

c. *Obtain equations that represent the new program.* Solve for the incoming variable from the limiting equation, and substitute it in the remaining equations of the current program. The new equations represent the revised program.

d. *Obtain the associated objective function.* Substitute the limiting equation (from step 3b) in the objective function associated with the preceding program. The result is the objective function associated with the revised program.

STEP 4. Test for optimality

a. If there is no positive-coefficient term in the modified objective function, the problem is solved.

b. Otherwise, the program should be revised by bringing in the largest positive-coefficient variable included in the modified objective function.

STEP 5.

Repeat steps 3b, 3c, 3d, and 4 until an optimal program has been designed. An optimal solution has been found when all coefficients in the modified objective function (step 4a) are negative.

A schematic diagram of this procedure is shown in Figure 4.1.

4.6 SUMMARY

We discussed the algebraic approach in this chapter for two principal reasons. First, in our progression from the graphical method, which has a very limited scope, to the more general and powerful simplex method, the algebraic method serves as an intermediate stage. Second, a familiarity with the algebraic approach is helpful in getting used to linear programming terminology and the structure of different solution stages as we search for the optimal solution. For example, it was stated that the algebraic method tests only the corner-point solutions. Further, as we proceed from one solution to another, the method exchanges one nonbasic (i.e., incoming) variable with one basic (i.e., outgoing) variable.* For each program we have a set of equations, and a modified objective function can be derived to test the optimality of a program.

Although more general than the graphical approach, the algebraic method is rather cumbersome for solving problems of large size. The algebraic method and the vector method are the foundations for the simplex method, which is one of the most general and powerful methods of solving linear programming problems. To appreciate the vector method and understand the mechanics of the simplex method, we need some background in matrix algebra. In Chapter 5 we present some basic concepts of matrix algebra that are sufficient for grasping the material discussed in this book.

---

* Sometimes, more than one basic variable will be "kicked out" of the program in this exchange process.

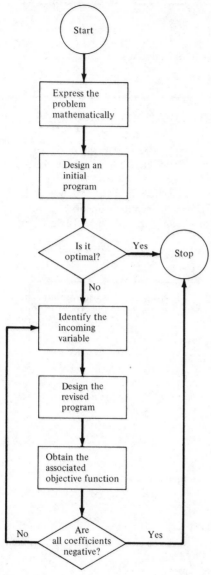

Figure 4.1 A Schematic of the Algebraic Method (Maximization Case)

REFERENCES

See references at the end of Chapter 3.

REVIEW QUESTIONS AND PROBLEMS

4.1. Solve Problem 3.4 by the algebraic method.

4.2. What action is suggested by the signs of the coefficients of the variables (candidates) in the modified (or associated) objective function at different solution stages? How would you relate the coefficients of the variables in the final objective function to the "marginal" concept in economic theory?

4.3. Solve the linear programming problem of Table 4.3 by the algebraic method.

Table 4.3

Resource	Processing Requirement for Product		Capacity Constraint
	A	B	
I	4	10	200
II	5	2	300
Profit contribution per unit	$60	$100	

4.4. Solve the following linear programming problem by the algebraic method:

Maximize $4X + 6Y$

subject to the constraints

$$2X + 3Y \leqslant 60$$
$$4X + Y \leqslant 40$$
$$X, \quad Y \geqslant 0$$

Give physical meaning to the algebraic statement of this problem.

4.5. Does the procedure for minimization differ from the one for maximization? If so, why? (See Section 7.5.)

5

Basic Concepts
of Matrix Algebra

5.1   INTRODUCTION

As described in previous chapters, a linear programming problem consists
of a linear objective function, a set of linear inequalities (or equations), and
a set of nonnegativity constraints. The set or system of equations that results
after the addition of slack or surplus variables is always such that it has more
unknowns (variables or activities) than the number of equations. The set
needs to be solved and, because it has more unknowns than the number of
equations, it can yield an infinite number of solutions. Of these, there is
usually one solution that is best or optimal.

Various methods of solving linear programming problems are essentially
systematic search procedures that test the optimality of selected* solutions.
The idea can be grasped graphically (Chapter 3), algebraically (Chapter 4),
or by the simplex method that utilizes a branch of mathematics known as
matrix algebra. Matrices and vectors are the central concepts of matrix
algebra, and they can be used effectively in explaining the mechanics of the
simplex method that we present in Chapter 7. The vector method, to be
described in Chapter 6, is actually nothing more than an explanation of the
simplex algorithm† in terms of vectors. It is, therefore, desirable that we
become familiar with matrices and vectors, their structure and properties,
and certain operations that are performed during the search process that
characterizes the simplex method. This chapter is devoted to this task.

---

* These are the basic feasible or extreme-point solutions. See Section 3.3.3. See also
Figure 5.7.
† An *algorithm* is a systematic search procedure.

99

5.2 MATRICES

5.2.1 Definition and Notation

A *matrix* is a rectangular array of ordered* numbers, consisting of m rows and n columns. The purpose of a matrix is to convey information in a concise fashion and lend ease of mathematical manipulation. Although a given matrix does not imply any mathematical operation, matrix algebra is a powerful tool for solving a system of linear equations.

Some examples of matrices are

$$\mathbf{A} = \begin{bmatrix} 2 & -1 \\ 1 & 3 \end{bmatrix} \quad \mathbf{B} = \begin{bmatrix} 2 & 1 & 4 \\ -3 & 2 & 1 \end{bmatrix} \quad \mathbf{C} = \begin{bmatrix} 10 & 6 & 2 \\ 5 & 10 & 5 \\ 1 & 2 & 2 \end{bmatrix}$$

Square brackets, [ ], will be used in this book to denote a matrix.† Any matrix in which the number of rows equals the number of columns is called a *square matrix*. Thus matrices $\mathbf{A}$ and $\mathbf{C}$ here are square matrices, whereas $\mathbf{B}$ is a rectangular matrix of two rows and three columns.

In general, a matrix $\mathbf{A}$ having m rows and n columns is written as

$$\mathbf{A} = \begin{bmatrix} a_{11} & a_{12} & \cdots & a_{1n} \\ a_{21} & a_{22} & \cdots & a_{2n} \\ \vdots & & \cdots & \vdots \\ a_{m1} & a_{m2} & \cdots & a_{mn} \end{bmatrix}$$

In this chapter, we shall use boldface capital letters, such as $\mathbf{A}$, $\mathbf{B}$, and $\mathbf{C}$, to denote the entire matrix, and lower case italic letters with proper subscripts, such as $a_{11}, a_{12}, \ldots$, to denote the numbers within the matrix.

Note that matrix $\mathbf{C}$ consists of ordered numbers that reflect the technical specification of the linear programming problem described in Section 1.2.

5.2.2 The Dimension of a Matrix

The number of rows and the number of columns in a given matrix determine the *dimension* or *order* of the matrix. For example, consider

$$\mathbf{D} = \begin{bmatrix} 2 & 1 \\ -1 & 4 \end{bmatrix} \quad \text{and} \quad \mathbf{E} = \begin{bmatrix} -3 & 2 & 1 \\ 2 & 1 & 3 \end{bmatrix}$$

* The term *ordered* implies that the position of each number is significant and must be determined carefully to represent the information contained in the problem.

† Sometimes matrices are denoted by parentheses, ( ), or by pairs of double vertical lines, ‖ ‖.

Matrix D is a 2×2 (two by two) matrix, whereas matrix E is a 2×3 matrix. When specifying the order or dimension of a matrix, the first number always refers to the rows of the matrix, the second number to the columns of the matrix. For example, the dimension of a matrix with m rows and n columns is $m \times n$. Rows of a matrix are numbered from top to bottom; columns are numbered from left to right.

5.2.3   Components of a Matrix

The various numbers within a matrix are referred to as the *components* of the matrix. For example,

$$\text{matrix } A = \begin{bmatrix} 2 & -1 \\ 1 & 3 \end{bmatrix}$$

has four components: 2, -1, 1, and 3. The general form of a 2×2 matrix is

$$A = \begin{bmatrix} a_{11} & a_{12} \\ a_{21} & a_{22} \end{bmatrix}$$

The components of the matrix are denoted by double subscripts. In the component a_{12}, the first subscript refers to the row and the second subscript refers to the column. The double subscripts give us the *address* of the component, indicating the specific row and column in which the component may be found. For example, component a_{ij} is located in the ith row and jth column.* A compact notation for a matrix is $A = [a_{ij}]$.

5.2.4   The Real Matrix

If all the components of a given matrix are real numbers, the matrix is called a *real matrix*. In this book we shall be dealing only with real matrices.

5.2.5   Some Special Matrices

Identity Matrix

The *identity matrix* (sometimes called the *unit matrix*) is a square matrix and is denoted by I. It is characterized by the fact that all components on its main diagonal (the diagonal going from the "northwest" corner to the "southeast" corner) are 1's whereas all other components are zero. Following are two different identity matrices:

$$I = \begin{bmatrix} 1 & 0 \\ 0 & 1 \end{bmatrix} \qquad I = \begin{bmatrix} 1 & 0 & 0 \\ 0 & 1 & 0 \\ 0 & 0 & 1 \end{bmatrix}$$

---

* For the sake of uniformity, i's will refer to rows and j's to columns throughout this book.

In other words, the identity matrix may be defined as follows:

$$\mathbf{I} = [a_{ij}] \qquad \text{where } a_{ij} = \begin{cases} 1 & \text{when } i = j \\ 0 & \text{when } i \neq j \end{cases}$$

The role of the identity matrix in matrix algebra is very similar to that played by the number 1 in ordinary algebra. Provided that they are compatible for multiplication, an identity matrix multiplied by any matrix gives the same matrix.* That is,

$$\mathbf{IA} = \mathbf{A}$$

and

$$\mathbf{AI} = \mathbf{A}$$

The Zero Matrix

The *zero matrix* is a matrix in which all elements are zero. It is denoted as **0**. Given next are three examples of zero matrices;

$$\mathbf{0} = \begin{bmatrix} 0 & 0 & 0 \\ 0 & 0 & 0 \\ 0 & 0 & 0 \end{bmatrix} \qquad \mathbf{0} = \begin{bmatrix} 0 & 0 \\ 0 & 0 \end{bmatrix} \qquad \mathbf{0} = \begin{bmatrix} 0 & 0 & 0 \\ 0 & 0 & 0 \end{bmatrix}$$

The role of the zero matrix in matrix algebra is very similar to that of zero in ordinary algebra. Provided that they are compatible for multiplication, the product of any matrix and the zero matrix is a zero matrix.

Transpose of a Matrix

Associated with every $m \times n$ matrix **A** is an $n \times m$ matrix (denoted by $\mathbf{A}^T$) whose rows are the columns of the given matrix **A**, in exactly the same order. In other words, the first row of **A** becomes the first column in the derived matrix $\mathbf{A}^T$, the second row becomes the second column, and so on. This derived matrix is called the *transpose* of **A**. Obviously, the transpose of the transpose matrix is the original matrix. For example, let

$$\mathbf{A} = \begin{bmatrix} 2 & 4 & -3 \\ 1 & 2 & 6 \\ 0 & 1 & 5 \end{bmatrix}$$

Then

$$\mathbf{A}^T = \begin{bmatrix} 2 & 1 & 0 \\ 4 & 2 & 1 \\ -3 & 6 & 5 \end{bmatrix}$$

* As will be explained later, two given matrices are compatible for multiplication only when the number of columns in the *lead matrix* equals the number of rows in the *lag matrix*.

and we can verify that the transpose of A^T gives the original matrix A.

$$[A^T]^T = A$$

5.3 VECTORS

5.3.1 Definition and Notation

A *vector* is an array of ordered numbers, consisting of a *single* row or column. If a vector has n components, it is an *n-dimensional vector* and corresponds to a point in an n-dimensional space. For example, a vector with just two components, 4 and 2, is a two-dimensional vector and corresponds to the point (4, 2). The same vector can also be graphed as line OV_3, which has *magnitude* as well as *direction** (see Figure 5.2).

The concept of a vector as a point in space is important because we can think of the right-hand side of a linear programming constraint set (e.g., the capacity levels of the three departments in Table 1.1) as a vector that corresponds to a point in space; and the solution of the linear programming problem can be thought of as reaching this point in space. We shall illustrate this vector approach in Chapter 6.

Like matrices, vectors will be denoted in this book by square brackets. We shall use boldface capital letters such as U and V to denote the entire vector, and lowercase italic letters with proper subscripts to denote the components of a vector.

In matrix algebra, vectors can be considered as special cases of a matrix. Two types of vectors can be identified: (1) row vectors, and (2) column vectors. A *row vector* is an ordered array of numbers arranged in a row. Following are examples of row vectors:

$$V_1 = [10 \quad 6 \quad 2] \qquad V_2 = [5 \quad 10 \quad 5] \qquad V_3 = [23 \quad 32 \quad 18]$$

Since it is a special case of a matrix, we can say that V_1 is a 1×3 matrix. Thus, in general, a row vector is a $1 \times n$ matrix, where $n = 1, 2, 3, \ldots$. A *column vector* is an ordered array of numbers arranged in a column. Examples of column vectors are

$$U_1 = \begin{bmatrix} 10 \\ 5 \\ 1 \end{bmatrix} \qquad U_2 = \begin{bmatrix} 6 \\ 10 \\ 2 \end{bmatrix} \qquad U_3 = \begin{bmatrix} 2{,}500 \\ 2{,}000 \\ 500 \end{bmatrix}$$

Since it is a special case of a matrix, we can say that U_1 is a 3×1 matrix. Thus, in general, a column vector is an $m \times 1$ matrix, where $m = 1, 2, 3, \ldots$.

---

* It should be emphasized that a vector is *not* a number. A *scalar*, which is a number, is distinguished from a vector by the fact that a scalar possesses only magnitude.

The numbers in a vector are referred to as the *components* of the vector. For example, the column vector U_1 has three components: 10, 5, and 1.

Note that row vector V_3 reflects the profit coefficients and the column vector U_3 the capacity levels of the linear programming problem presented in Section 1.2.

5.3.2   Some Special Vectors

Unit Vector

A *unit vector* is a vector in which one component has the value 1 while the rest of the components are zeros. Here are some examples of unit vectors:

$$U_1 = \begin{bmatrix} 1 \\ 0 \\ 0 \end{bmatrix} \qquad U_2 = \begin{bmatrix} 0 \\ 1 \\ 0 \end{bmatrix} \qquad V_1 = [1 \quad 0 \quad 0] \qquad V_2 = [0 \quad 1 \quad 0]$$

Zero Vector

A *zero vector* is a vector in which all the components are zero. Given next are a 1×3 zero row vector and a 3×1 zero column vector:

$$\mathbf{0} = [0 \quad 0 \quad 0] \qquad \mathbf{0} = \begin{bmatrix} 0 \\ 0 \\ 0 \end{bmatrix}$$

Transpose of a Vector

Consider a $1 \times m$ row vector:

$$V = [a_1 \quad a_2 \quad \cdots \quad a_m]$$

If we write this vector vertically with the same components in exactly the same order, we obtain the transpose of V:

$$V^T = U = \begin{bmatrix} a_1 \\ a_2 \\ \vdots \\ a_m \end{bmatrix}$$

The column vector U, then, is the transpose of the row vector V. Obviously, the given row vector V is the transpose of the column vector U. That is,

$$U^T = V \qquad \text{or} \qquad [V^T]^T = V$$

Linearly Independent Vectors

A set of vectors of the same dimension is *linearly independent* if any one vector in the set cannot be expressed as a linear combination of the remaining vectors in the set. It is obvious from this definition that a set consisting of unit vectors is always linearly independent. A precise definition of linear independence is given in Section 5.12.

EXAMPLE 5.1

The set V_1, V_2 is linearly independent if

$$V_1 = [1 \quad 0] \quad \text{and} \quad V_2 = [0 \quad 1]$$

EXAMPLE 5.2

The set U_1, U_2, U_3 is linearly independent if

$$U_1 = \begin{bmatrix} 1 \\ 0 \\ 0 \end{bmatrix} \quad U_2 = \begin{bmatrix} 0 \\ 1 \\ 0 \end{bmatrix} \quad U_3 = \begin{bmatrix} 0 \\ 0 \\ 1 \end{bmatrix}$$

EXAMPLE 5.3

The set V_1, V_2 is *not* linearly independent if

$$V_1 = [2 \quad 4] \quad \text{and} \quad V_2 = [4 \quad 8]$$

Note that in Example 5.3, V_2 can simply be obtained by multiplying each component of V_1 by the real number 2.

5.4 GRAPHICAL REPRESENTATION OF VECTORS

If we assume that the vectors emanate from the origin of a coordinate system, it is easy to view a vector as a point in space, and vice versa. A given vector can be represented graphically if it has less than four components. Consider, for example, a vector $V_1 = [5]$, which has a single component. This vector can be represented in a one-dimensional space as in Figure 5.1. Similarly, a vector $V_2 = [-3]$ can be represented in a one-dimensional space as in Figure 5.1.

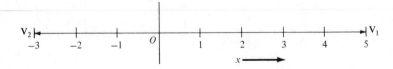

Figure 5.1

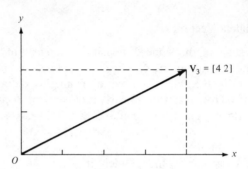

Figure 5.2

A vector having two components can be represented in a two-dimensional space. For example, the vector $\mathbf{V}_3 = [4 \quad 2]$ can be graphed as shown in Figure 5.2. That is, $\mathbf{V}_3 = [4 \quad 2]$ implies that $X = 4$ and $Y = 2$. Similarly, a three-component vector can be represented in a three-dimensional space. For example, the vector

$$\mathbf{V}_4 = [2 \quad 1 \quad 4]$$

is graphed in Figure 5.3. That is,

$$\mathbf{V}_4 = [2 \quad 1 \quad 4]$$

implies that $X = 2$, $Y = 1$, and $Z = 4$.

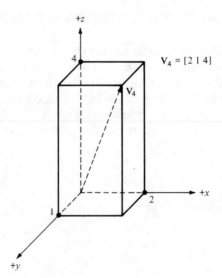

Figure 5.3

We note that there is no geometric distinction between row and column vectors. Figure 5.3, for example, is a graphical representation of

$$\mathbf{V}_4 = [2 \quad 1 \quad 4] \quad \text{as well as of} \quad \mathbf{V}_5 = \begin{bmatrix} 2 \\ 1 \\ 4 \end{bmatrix}$$

In general, it takes an m-dimensional space to represent an m-component vector. Evidently, we are limited by our inability to graph a space having more than three dimensions. However, the concept of correspondence between the number of components in a vector and the number of dimensions required to represent it is very important.

5.5 VECTOR NOTATION OF A MATRIX

If we consider vectors as special cases of a matrix, a given matrix can always be represented as a set of row or column vectors. For example, let $\mathbf{A}$ be a 3×3 matrix:

$$\mathbf{A} = \begin{bmatrix} 10 & 6 & 2 \\ 5 & 10 & 5 \\ 1 & 2 & 2 \end{bmatrix}$$

The matrix $\mathbf{A}$, when represented as a set of three row vectors, $\mathbf{V}_1, \mathbf{V}_2, \mathbf{V}_3$, can be written

$$\mathbf{A} = \begin{bmatrix} \mathbf{V}_1 \\ \mathbf{V}_2 \\ \mathbf{V}_3 \end{bmatrix} \quad \text{where} \begin{matrix} \mathbf{V}_1 = [10 \quad 6 \quad 2] \\ \mathbf{V}_2 = [\ 5 \quad 10 \quad 5] \\ \mathbf{V}_3 = [\ 1 \quad 2 \quad 2] \end{matrix}$$

Note that row vectors $\mathbf{V}_1, \mathbf{V}_2, \mathbf{V}_3$ reflect the input–output coefficients of the linear programming problem of Section 1.2.

The same matrix $\mathbf{A}$, when represented as a set of three column vectors $\mathbf{U}_1, \mathbf{U}_2, \mathbf{U}_3$, can be written

$$\mathbf{A} = [\mathbf{U}_1 \quad \mathbf{U}_2 \quad \mathbf{U}_3]$$

where

$$\mathbf{U}_1 = \begin{bmatrix} 10 \\ 5 \\ 1 \end{bmatrix} \quad \mathbf{U}_2 = \begin{bmatrix} 6 \\ 10 \\ 2 \end{bmatrix} \quad \mathbf{U}_3 = \begin{bmatrix} 2 \\ 5 \\ 2 \end{bmatrix}$$

Note that column vectors $\mathbf{U}_1, \mathbf{U}_2, \mathbf{U}_3$ also represent the input–output coefficients of the linear programming problem of Section 1.2.*

* The row and column arrangement of input–output coefficients lends importance to the term *ordered*. The two arrangements order the same numbers to convey information with different emphasis.

5.6 BASIC CONCEPTS AND OPERATIONS CONCERNING MATRICES AND VECTORS

Since vectors can be considered as special cases of matrices, the rules of operations involving vectors and matrices and their behavior are similar.

5.6.1 Equality of Matrices

Two matrices are equal if and only if (1) their dimension or order is the same, and (2) their corresponding components are equal to each other. Thus, if

$$A = \begin{bmatrix} 2 & 1 \\ -1 & 3 \\ 4 & 2 \end{bmatrix} \quad B = \begin{bmatrix} 2 & -1 & 4 \\ 1 & 3 & 2 \end{bmatrix} \quad C = \begin{bmatrix} 2 & -1 & 4 \\ 2 & 3 & 2 \end{bmatrix}$$

then

$\quad\quad$ $A \neq B$ $\quad$ because their dimensions are not the same

$\quad\quad$ $B \neq C$ $\quad$ because their components are not the same

If A and B have the same dimensions, and $a_{ij} = b_{ij}$ for all i and j, then $A = B$. Conversely, if two matrices are equal, their corresponding components are equal.

5.6.2 Equality of Vectors

Two given vectors are said to be equal if and only if (1) they are the same type of vectors, and (2) their corresponding components are equal to each other. Thus, if

$$U = [2 \quad 1 \quad 0] \quad V = [2 \quad 1 \quad 0] \quad X = \begin{bmatrix} 2 \\ 1 \\ 0 \end{bmatrix}$$

$$W = [2 \quad -1 \quad 0] \quad Y = \begin{bmatrix} 2 \\ 1 \\ 1 \end{bmatrix} \quad Z = \begin{bmatrix} 2 \\ 1 \\ 0 \end{bmatrix}$$

then

$$U = V \quad\quad U \neq W$$
$$X = Z \quad\quad X \neq Y$$
$$U \neq X \quad\quad U \neq Y$$

5.6.3 Addition and Subtraction of Matrices

Two matrices **A** and **B** can be added *only* if they have the same dimensions. Similarly, a matrix **B** can be subtracted from matrix **A** only if both **A** and **B** have the same dimensions. Once it is established that the numbers of rows and columns of the two matrices are identical, their respective components can be added together (or subtracted from each other).

EXAMPLE 5.4

Let

$$A = \begin{bmatrix} 2 & 3 & 4 \\ 1 & 0 & 6 \end{bmatrix} \quad B = \begin{bmatrix} -1 & 2 & 1 \\ 0 & 3 & 2 \end{bmatrix}$$

Then

$$A + B = \begin{bmatrix} 2 & 3 & 4 \\ 1 & 0 & 6 \end{bmatrix} + \begin{bmatrix} -1 & 2 & 1 \\ 0 & 3 & 2 \end{bmatrix} = \begin{bmatrix} 1 & 5 & 5 \\ 1 & 3 & 8 \end{bmatrix}$$

Two things must be observed in matrix addition. First, the matrix representing the sum of **A** and **B** has the same dimension as **A** and **B**. Second, the order of addition is not important, for if **A** + **B** equals **C**, then **B** + **A** also equals **C**.

Since the order of addition is not important in matrix addition, we say that matrix addition obeys the commutative law of addition. In this sense, matrix addition is similar to the addition of numbers in ordinary algebra.

5.6.4 Addition and Subtraction of Vectors

The addition and subtraction of vectors, as in the case of matrices, is defined only if both vectors have the same dimensions. This implies that two given vectors can be added (or subtracted from one another) *only* if (1) they are the same type of vectors, and (2) they have the same number of components.

EXAMPLE 5.5

Let

$$V_1 = [2 \quad 3 \quad 4] \quad V_2 = [-1 \quad 2 \quad 1]$$

Then

$$V_1 + V_2 = [1 \quad 5 \quad 5]$$

Let

$$U_1 = \begin{bmatrix} 2 \\ 1 \end{bmatrix} \quad U_2 = \begin{bmatrix} -1 \\ 0 \end{bmatrix}$$

Then

$$U_1 + U_2 = \begin{bmatrix} 1 \\ 1 \end{bmatrix}$$

As in the case of matrices, the order of addition of vectors is not important.

5.6.5 Graphical Representation of the Addition of Vectors

Let us consider two row vectors $V_1 = [2 \quad 1]$ and $V_2 = [1 \quad 3]$. Their sum is $V_1 + V_2 = V_3 = [3 \quad 4]$. This type of addition can be represented graphically as shown in Figure 5.4. The vector V_3 has been obtained

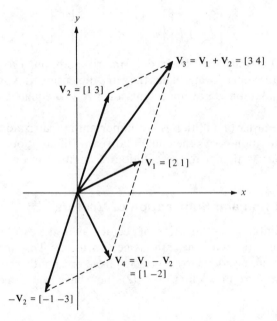

Figure 5.4

by adding the vectors V_1 and V_2. Two things should be observed: (1) Both V_1 and V_2 are two-component vectors and hence need a two-dimensional space to be represented graphically, and (2) the vector V_3 lies in the same two-dimensional space. Note that the vector V_3 is the diagonal, passing through the origin, of the parallelogram formed by the vectors V_1 and V_2. A graphical representation of $V_1 - V_2$ is also shown in Figure 5.4. $V_1 - V_2 = V_4 = [1 \quad -2]$. Note that $-V_2$ has the same magnitude as V_2 except that its direction has been completely reversed.

5.6.6 Scalar Multiplication and Linear Combination

The simultaneous multiplication of all the components of a given matrix by a real number (scalar) is called *scalar multiplication*. Suppose that an $m \times n$ matrix $\mathbf{A} = [a_{ij}]$ is to be multiplied by a scalar k. Then

$$\mathbf{A}k = k\mathbf{A} = [ka_{ij}]$$

EXAMPLE 5.6

Let

$$\mathbf{A} = \begin{bmatrix} -2 & 3 & 1 \\ 0 & 2 & 4 \\ 3 & -5 & 1 \end{bmatrix} \quad \text{then } 2\mathbf{A} = \begin{bmatrix} -4 & 6 & 2 \\ 0 & 4 & 8 \\ 6 & -10 & 2 \end{bmatrix}$$

Similarly, if we multiply all the elements of a given vector by some scalar (i.e., a real number), we get a scalar multiple of the given vector. In the case of vectors, scalar multiplication can be given a graphical interpretation (see Figure 5.5). Consider, for example, a vector $\mathbf{V}_1 = \begin{bmatrix} 1 & 2 \end{bmatrix}$. Then

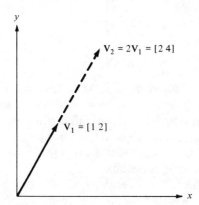

Figure 5.5

$2\mathbf{V}_1 = \mathbf{V}_2 = \begin{bmatrix} 2 & 4 \end{bmatrix}$. The vector $\mathbf{V}_2$ is graphed in Figure 5.5. As shown in Figure 5.5, the multiplication of a given vector by a positive scalar having a value greater than 1 has resulted in the elongation of the given vector without changing its direction. Clearly, the multiplication of a given vector by the scalar 1 will leave its magnitude unaltered, whereas multiplication by a positive scalar having a value less than 1 will result in a shortening of the given vector. Thus,

$$\text{if } k = 1: \quad \text{then } k\mathbf{V}_1 = \begin{bmatrix} 1 & 2 \end{bmatrix} = \mathbf{V}_1$$

$$\text{if } k = \tfrac{1}{2}: \quad \text{then } k\mathbf{V}_1 = \begin{bmatrix} \tfrac{1}{2} & 1 \end{bmatrix} = \tfrac{1}{2}\mathbf{V}_1$$

The multiplication of a given vector by a positive scalar, therefore, can change the length of the vector without affecting its direction.

Let us now multiply V_1 by the scalar -1. Then $-1V_1 = [-1 \quad -2] = -V_1$. The vector $-V_1$ is graphed in Figure 5.6. The result of multiplying V_1 by the scalar -1 has been, as shown in the figure, to reverse the direction of V_1 without affecting its magnitude.

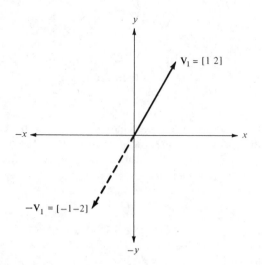

Figure 5.6

The illustrations of scalar multiplication (Figures 5.5 and 5.6) serve to indicate that the length as well as the direction of a given vector can be manipulated. Scalar multiplication, in conjunction with the operation of vector addition, gives us a very powerful tool for expressing a given vector as a linear combination of some other vectors. It is this capability that we use in solving linear programming problems.

Let us illustrate the power of scalar multiplication and vector addition by attempting to see whether we can represent any vector, of specified dimension, in a space of appropriate dimensions. Consider, first, a single-dimensional space. It is obvious that all vectors in this space can be expressed as scalar multiples of the vector $V_1 = [1]$. For example,

$$\text{if } V_2 = [-6], \quad \text{then } V_2 = -6V_1$$
$$\text{if } V_3 = [8], \quad \text{then } V_3 = 8V_1$$

Consider, next, a two-dimensional space. It is obvious that all vectors in the two-dimensional space can be expressed as the sum of scalar multiples of the vectors

$$U_1 = \begin{bmatrix} 1 \\ 0 \end{bmatrix} \quad \text{and} \quad U_2 = \begin{bmatrix} 0 \\ 1 \end{bmatrix}$$

For example,

$$\text{if } U_3 = \begin{bmatrix} -6 \\ 8 \end{bmatrix}, \quad \text{then } U_3 = -6U_1 + 8U_2$$

$$\text{if } U_4 = \begin{bmatrix} 10 \\ 12 \end{bmatrix}, \quad \text{then } U_4 = 10U_1 + 12U_2$$

The illustration in the two-dimensional space is an example of what is called *linear combination*. A linear combination is essentially a sum of the scalar multiples of some vectors. What interests us, however, is a linear combination that involves some special type of vectors, namely, linearly independent vectors.* The reason for our interest in a set of linearly independent vectors lies in the fact that by forming linear combinations of the vectors in this set, we can reach any point in a space of specified dimensions. For example, by forming a combination of two linearly independent (two-dimensional) vectors, we can reach any point in a two-dimensional space. Similarly, we need three linearly independent (three-dimensional) vectors in order to reach any point in a three-dimensional space. In general, we need m linearly independent (m-dimensional) vectors in order to reach any point in the m-dimensional space. That is, any m-dimensional vector can be expressed as a linear combination of m linearly independent (m-dimensional) vectors. This statement, as we shall discover in Chapter 6, is of extreme significance in solving a linear programming problem. The linear programming problem boils down to expressing an m-dimensional vector as a linear combination of a set of linearly independent vectors.

5.6.7   Multiplication of Vectors

Row Vector × Column Vector

We shall first define the product of a row vector and a column vector, both having the same number of components. Let

$$V_1 = [2 \quad 3 \quad 4] \qquad U_1 = \begin{bmatrix} 1 \\ -2 \\ 4 \end{bmatrix}$$

* The vectors

$$V_1 = \begin{bmatrix} 1 \\ 0 \end{bmatrix} \quad \text{and} \quad V_2 = \begin{bmatrix} 0 \\ 1 \end{bmatrix}$$

are linearly independent. A precise definition of linear independence is given in Section 5.12. For the time being, the reader can consider that a set of vectors is linearly independent if any one vector in the set cannot be formed by taking a linear combination of the other vectors in the set. For example, if $V_1 = [1 \quad 0 \quad 0]$, $V_2 = [0 \quad 1 \quad 0]$, and $V_3 = [0 \quad 0 \quad 1]$, then the vectors V_1, V_2, and V_3 are linearly independent. In the two-dimensional space, two vectors are independent if when graphed they do not lie on the same line. For example, the vectors $V_1 = [1 \quad 0]$ and $V_2 = [0 \quad 1]$ are linearly independent, but the vectors $V_3 = [2 \quad 4]$ and $V_4 = [4 \quad 8]$ are not linearly independent.

Then

$$\mathbf{V}_1 \times \mathbf{U}_1 = 2(1) + 3(-2) + 4(4) = 12$$

In general, the product of a $1 \times n$ row vector by an $n \times 1$ column vector is given by

$$[a_1 a_2 \cdots a_n] \begin{bmatrix} b_1 \\ b_2 \\ \vdots \\ b_n \end{bmatrix} = a_1 b_1 + a_2 b_2 + \cdots + a_n b_n = \sum_{i=1}^{n} a_i b_i \qquad (5.1)$$

It is to be observed that the number of components in the lead vector is exactly the same as the number of components in the lag vector. If this condition does not exist, the vectors are said to be *incompatible* and their multiplication is not defined. Furthermore, in any multiplication involving a row vector and a column vector, the product is treated as a scalar,* provided that the row vector is the lead vector. This scalar is the sum of the products of the corresponding components of the two vectors, as shown in Equation (5.1). Let us illustrate the multiplication operation and then check on dimensionality to ensure compatibility for multiplication. If

$$\mathbf{V}_1 = [2 \quad 3 \quad 4] \qquad \text{and} \qquad \mathbf{U}_1 = \begin{bmatrix} 1 \\ -2 \\ 4 \end{bmatrix}$$

then

$$\mathbf{V}_1 \mathbf{U}_1 = \quad \mathbf{V}_1 \quad \times \quad \mathbf{U}_1 \quad = \quad 12$$
$$\text{Dimensions:} \qquad (1 \times 3) \quad (3 \times 1) \quad (1 \times 1)$$

Note that the order 1×1 means that we have a single component.

The following dimensional arrangement must hold for compatibility in vector multiplication:

$$\text{lead vector} \times \text{lag vector} = \text{product}$$
$$\text{Dimensions:} \quad (1 \times n) \qquad (n \times 1) \qquad (1 \times 1)$$

Column Vector × Row Vector

Let

$$\mathbf{U}_1 = \begin{bmatrix} 2 \\ 1 \\ -3 \end{bmatrix} \qquad \mathbf{V}_1 = [1 \quad -2 \quad 3]$$

---

* Actually, the matrix product of a row and column vector yields a matrix having only one component. However, there is complete equivalence between scalars (real numbers) and matrices with only one component.

Then

$$\mathbf{U}_1 \times \mathbf{V}_1 = \begin{bmatrix} 2 \\ 1 \\ -3 \end{bmatrix} \begin{bmatrix} 1 & -2 & 3 \end{bmatrix} = \begin{bmatrix} 2 & -4 & 6 \\ 1 & -2 & 3 \\ -3 & 6 & -9 \end{bmatrix}$$

It is to be observed again that $\mathbf{U}_1$ and $\mathbf{V}_1$ are dimensionally compatible. However, the product in this case (when the column vector is the lead vector) is a 3×3 matrix.

In general, if the lead vector is an $n \times 1$ column vector and the lag vector is a $1 \times n$ row vector, their product results in an $n \times n$ matrix:

$$\text{lead vector} \times \text{lag vector} = \text{product}$$
Dimensions: $n \times 1$ $1 \times n$ $n \times n$

5.6.8 Multiplication of Matrices

The definition of multiplication of a row vector by a column vector can easily be extended to cover matrix multiplication. Let us illustrate matrix multiplication by considering a specific example. Let

$$\mathbf{A} = \begin{bmatrix} 2 & 1 & -2 \\ 3 & 2 & 4 \end{bmatrix} \qquad \mathbf{B} = \begin{bmatrix} 1 & 2 \\ 0 & 3 \\ -2 & 1 \end{bmatrix}$$

and let $\mathbf{AB} = \mathbf{C}$. Then

$$\mathbf{A} \times \mathbf{B} = \mathbf{C} = \begin{bmatrix} 2(1) + 1(0) + (-2)(-2) & 2(2) + 1(3) + (-2)(1) \\ 3(1) + 2(0) + 4(-2) & 3(2) + 2(3) + 4(1) \end{bmatrix}$$

or

$$\mathbf{C} = \begin{bmatrix} 6 & 5 \\ -5 & 16 \end{bmatrix}$$

This matrix multiplication consists of the following steps:

1. *Check on compatibility.* Is the number of columns in the lead matrix (**A**) equal to the number of rows in the lag matrix (**B**)? If so, the matrices are compatible for multiplication; otherwise, not. In the above case, the multiplication $\mathbf{A} \times \mathbf{B}$ is compatible.

2. *The operation of multiplication.* The components of the first row of **A**, the lead matrix, are multiplied by the corresponding components of the first column of **B**, the lag matrix. The product is summed and is placed in the first row–first column cell of the resultant matrix **C**. Similarly, the components of the second row of matrix **A** are multiplied by the corresponding components of the first column of matrix **B**; the product is summed and is placed in the second row–first column element of the resultant matrix, and

so on. The resultant matrix **C** is a 2×2 matrix. The check on dimensional compatibility is

$$\text{lead matrix} \times \text{lag matrix} = \text{resultant matrix}$$
$$\mathbf{A} \quad \times \quad \mathbf{B} \quad = \quad \mathbf{C}$$
$$\text{Dimensions:} \quad (2 \times 3) \quad (3 \times 2) \quad (2 \times 2)$$

In general, if we multiply an $m \times k$ matrix **A** by another $k \times n$ matrix **B**, the dimension of the resultant matrix is $m \times n$:

$$\text{lead matrix} \times \text{lag matrix} = \text{resultant matrix}$$
$$\text{Dimensions:} \quad (m \times k) \quad (k \times n) \quad (m \times n)$$

EXAMPLE 5.7

Let

$$\mathbf{C} = \begin{bmatrix} 4 & -1 & 2 \\ 0 & 2 & 3 \end{bmatrix} \quad \mathbf{D} = \begin{bmatrix} 2 & 3 & 1 \\ 0 & 1 & 0 \\ 2 & 1 & 0 \end{bmatrix}$$

Then

$$\mathbf{CD} = \begin{bmatrix} 12 & 13 & 4 \\ 6 & 5 & 0 \end{bmatrix}$$

but **DC** is incompatible.

EXAMPLE 5.8

Obtain the product **BA** when

$$\mathbf{A} = \begin{bmatrix} 2 & 1 & -2 \\ 3 & 2 & 4 \end{bmatrix} \quad \mathbf{B} = \begin{bmatrix} 1 & 2 \\ 0 & 3 \\ -2 & 1 \end{bmatrix}$$

$$\mathbf{BA} = \begin{bmatrix} 1 & 2 \\ 0 & 3 \\ -2 & 1 \end{bmatrix} \begin{bmatrix} 2 & 1 & -2 \\ 3 & 2 & 4 \end{bmatrix}$$

$$= \begin{bmatrix} 1(2) + 2(3) & 1(1) + 2(2) & 1(-2) + 2(4) \\ 0(2) + 3(3) & 0(1) + 3(2) & 0(-2) + 3(4) \\ -2(2) + 1(3) & -2(1) + 1(2) & -2(-2) + 1(4) \end{bmatrix}$$

$$= \begin{bmatrix} 8 & 5 & 6 \\ 9 & 6 & 12 \\ -1 & 0 & 8 \end{bmatrix}$$

In this example, **AB** and **BA** are both compatible, but note that $\mathbf{AB} \neq \mathbf{BA}$.

Table 5.1

Law	Ordinary Algebra	Matrix Algebra
1a. *Commutative law of addition* (dealing with the order of operations). Comment: Matrix addition obeys the commutative law of addition.	$a + b = b + a$	$\mathbf{A} + \mathbf{B} = \mathbf{B} + \mathbf{A}$
1b. *Commutative law of multiplication.* Comment: Matrix multiplication, in general, does not obey the commutative law of multiplication.	$a \times b = b \times a$	$\mathbf{AB} \neq \mathbf{BA}$ (except in certain cases and where either $\mathbf{A}$ or $\mathbf{B}$ is an identity matrix)
2a. *Associative law of addition* (order remains the same; deals with the sequence of similar operations within a given order). Comment: Matrix addition obeys the associative law of addition.	$a + (b + c) = (a + b) + c$	$\mathbf{A} + (\mathbf{B} + \mathbf{C}) = (\mathbf{A} + \mathbf{B}) + \mathbf{C}$
2b. *Associative law of multiplication.* Comment: Matrix multiplication obeys the associative law of multiplication.	$a(bc) = (ab)c$	$\mathbf{A}(\mathbf{BC}) = (\mathbf{AB})\mathbf{C}$
3. *Distributive law* (deals with the sequence of addition and multiplication operations within a given order). Comment: Matrix algebra obeys the distributive law.	$a(b + c) = ab + ac$ $(d + e)f = df + ef$	$\mathbf{A}(\mathbf{B} + \mathbf{C}) = \mathbf{AB} + \mathbf{AC}$ $(\mathbf{D} + \mathbf{E})\mathbf{F} = \mathbf{DF} + \mathbf{EF}$

This leads to the remark that in matrix multiplication *order is important*. Matrix multiplication, therefore, does not obey the commutative law of multiplication.

5.6.9 Multiplication of a Matrix by a Vector

The multiplication of a matrix by a vector follows the rules of regular matrix multiplication, as explained previously.

EXAMPLE 5.9

Let

$$\mathbf{A} = \begin{bmatrix} 4 & -1 & 2 \\ 0 & 2 & 3 \end{bmatrix} \quad \text{and} \quad \mathbf{U} = \begin{bmatrix} 1 \\ 0 \\ 2 \end{bmatrix}$$

Then

$$\mathbf{AU} = \begin{bmatrix} 4 & -1 & 2 \\ 0 & 2 & 3 \end{bmatrix} \begin{bmatrix} 1 \\ 0 \\ 2 \end{bmatrix} = \begin{bmatrix} 8 \\ 6 \end{bmatrix}$$

Comparisons between various operations in ordinary algebra and matrix algebra are shown in Table 5.1.

5.7 DETERMINANTS

5.7.1 Definition and Notation

Associated with any square matrix is a number that is called its *determinant*. Consider, for example, a square matrix

$$\mathbf{A} = \begin{bmatrix} 2 & 1 & 3 \\ -1 & 4 & 1 \\ 3 & 2 & 5 \end{bmatrix}$$

This square matrix has a determinant that, in its unexpanded form, is written as follows:

$$|\mathbf{A}| = \begin{vmatrix} 2 & 1 & 3 \\ -1 & 4 & 1 \\ 3 & 2 & 5 \end{vmatrix}$$

The reader should note that the determinant is denoted by a pair of single vertical bars of the form | |, while the matrix notation was [ ]. Further, whereas the given matrix **A** did not imply any mathematical operation, the determinant of this matrix (having exactly the same components in the same positions) does imply certain operations. When the determinant appears in the above form, it is said to be in its *unexpanded* form. To evaluate a determinant, we expand it according to certain rules and obtain a single number. Thus it is always possible to calculate the numerical value of a determinant.

The concept of determinants is very useful. By evaluating the determinant of a given set of linear equations, for example, one can immediately determine whether or not the set has a unique solution. If the given set does not have a unique solution, further tests (also using the concept of determinants) will show that the set either has no solution or has an infinite number of solutions. Furthermore, if a given set of linear equations has a unique solution, its determinant is used to find the values of the unknowns in that set.*

5.7.2 How to Calculate the Numerical Value of Determinants

Before giving a generalized method of evaluating determinants, we shall show how a 2×2 determinant is evaluated. Consider a matrix

$$\mathbf{A} = \begin{bmatrix} 2 & 4 \\ 3 & 7 \end{bmatrix}$$

The determinant of **A** is

$$|\mathbf{A}| = \begin{vmatrix} 2 & 4 \\ 3 & 7 \end{vmatrix} = 14 - 12 = 2$$

Here the main diagonal is $2 \to 7$; and the secondary diagonal is $3 \to 4$. Note, then, that the numerical value of a 2×2 determinant is calculated by taking the product of its main diagonal components and subtracting from it the product of the secondary diagonal components.

In general, if there is a 2×2 matrix

$$\mathbf{A} = \begin{bmatrix} a_{11} & a_{12} \\ a_{21} & a_{22} \end{bmatrix}$$

then its determinant

$$|\mathbf{A}| = \begin{vmatrix} a_{11} & a_{12} \\ a_{21} & a_{22} \end{vmatrix} = a_{11}a_{22} - a_{21}a_{12}$$

Before discussing the procedure for calculating the numerical value of determinants of order higher than 2×2, we present the definitions of a minor and a cofactor.

---

* See Appendices B, C, and D.

Minor

The determinant of the submatrix formed by deleting one row and one column from a given square matrix is called a *minor*. Consider, for example, the matrix

$$\mathbf{A} = \begin{bmatrix} 2 & 4 & 1 \\ 3 & 7 & 2 \\ 1 & 0 & -4 \end{bmatrix}$$

If we delete the first row and the first column, we get the submatrix

$$\mathbf{A'} = \begin{bmatrix} 7 & 2 \\ 0 & -4 \end{bmatrix}$$

The determinant of $\mathbf{A'}$ is

$$|\mathbf{A'}| = \begin{vmatrix} 7 & 2 \\ 0 & -4 \end{vmatrix}$$

The determinant $|\mathbf{A'}|$, as defined above, is the minor obtained by deleting the first row and the first column of the given matrix. This minor is denoted by M_{11}. The first subscript of the minor always refers to the row being deleted from the matrix, and the second subscript refers to the column being deleted. Thus

$$M_{11} = \begin{vmatrix} 7 & 2 \\ 0 & -4 \end{vmatrix} = 7(-4) - (0)(2) = -28$$

Similarly, M_{12} is the minor obtained by deleting the first row and the second column. In our example,

$$M_{12} = \begin{vmatrix} 3 & 2 \\ 1 & -4 \end{vmatrix} = 3(-4) - (1)(2) = -14$$

In general, M_{ij} is the minor obtained by deleting the ith row and the jth column of a given matrix.*

Cofactor

Associated with each component a_{ij} of a square matrix is a cofactor denoted by C_{ij}. The subscripts i and j refer to the row and the column in which the component is located. Consider, for example, the matrix

$$\mathbf{A} = \begin{bmatrix} 2 & 4 & 1 \\ 3 & 7 & 2 \\ 1 & 0 & -4 \end{bmatrix}$$

* For further discussion on minors, see Loomba and Turban [1974, pp. 32–34].

The cofactor of the component 3 (which is located in the second row and first column) is denoted by C_{21} and is defined as

$$C_{21} = (-1)^{2+1} \begin{vmatrix} 4 & 1 \\ 0 & -4 \end{vmatrix}$$

But we already know that the determinant $\begin{vmatrix} 4 & 1 \\ 0 & -4 \end{vmatrix}$ is M_{21}, that is, the minor obtained by deleting the second row and first column of matrix **A**. The only difference between C_{21} and M_{21}, therefore, is the sign generated by the expression $(-1)^{2+1}$. Indeed, this is exactly what differentiates a cofactor from its minor. A cofactor is a minor with its proper sign, and this sign is determined by the subscripts of the component with which the cofactor is associated. Consider, for example, the matrix

$$\mathbf{A} = \begin{bmatrix} a_{11} & a_{12} & a_{13} \\ a_{21} & a_{22} & a_{23} \\ a_{31} & a_{32} & a_{33} \end{bmatrix}$$

The cofactor of a_{11} is

$$(-1)^{1+1} \begin{vmatrix} a_{22} & a_{23} \\ a_{32} & a_{33} \end{vmatrix}$$

In general, the cofactor of any component a_{ij} (where i refers to the row and j to the column) is denoted by C_{ij}, and its value is given by $C_{ij} = (-1)^{i+j} M_{ij}$. Note that $(-1)^{i+j}$ is positive when $(i + j)$ is even and negative when $(i + j)$ is odd.

Calculating the Numerical Value of Determinants by Use of Cofactors

We state here, without proof, that the numerical value of any determinant is equal to the sum of the products of components of any row (or column) and their corresponding cofactors with the same subscripts. We shall refer to this method as "expansion of a determinant by cofactors." This method is general and can be applied to calculate the value of a determinant of any size. We apply the definition to a 3 × 3 matrix.

$$\mathbf{A} = \begin{bmatrix} a_{11} & a_{12} & a_{13} \\ a_{21} & a_{22} & a_{23} \\ a_{31} & a_{32} & a_{33} \end{bmatrix}$$

The determinant of **A** can be calculated by expanding it along any row or column as follows:

$$|\mathbf{A}| = a_{11}C_{11} + a_{12}C_{12} + a_{13}C_{13}$$ determinant expanded along first row

or

$$|\mathbf{A}| = a_{21}C_{21} + a_{22}C_{22} + a_{23}C_{23}$$ determinant expanded along second row

Similarly,

$$|\mathbf{A}| = a_{11}C_{11} + a_{21}C_{21} + a_{31}C_{31}$$ determinant expanded along first column

$$|\mathbf{A}| = a_{13}C_{13} + a_{23}C_{23} + a_{33}C_{33}$$ determinant expanded along third column

Note that the subscripts of the component and its cofactor are the same in each term of the expansion.

EXAMPLE 5.10

Consider a matrix

$$\mathbf{A} = \begin{bmatrix} 2 & 3 & 2 \\ -1 & 0 & 1 \\ 2 & 1 & 3 \end{bmatrix}$$

The determinant of **A** is

$$|\mathbf{A}| = \begin{vmatrix} 2 & 3 & 2 \\ -1 & 0 & 1 \\ 2 & 1 & 3 \end{vmatrix}$$

Expanding along the second row, we obtain

$$|\mathbf{A}| = a_{21}C_{21} + a_{22}C_{22} + a_{23}C_{23}$$

$$= (-1)(-1)^{2+1}\begin{vmatrix} 3 & 2 \\ 1 & 3 \end{vmatrix} + 0C_{22} + 1(-1)^{2+3}\begin{vmatrix} 2 & 3 \\ 2 & 1 \end{vmatrix}$$

$$= (9 - 2) + 0 + (-1)(2 - 6)$$

$$= 11$$

Expanding along the second column, we obtain

$$|\mathbf{A}| = a_{12}C_{12} + a_{22}C_{22} + a_{32}C_{32}$$

$$= 3(-1)^{1+2}\begin{vmatrix} -1 & 1 \\ 2 & 3 \end{vmatrix} + 0C_{22} + 1(-1)^{3+2}\begin{vmatrix} 2 & 2 \\ -1 & 1 \end{vmatrix}$$

$$= 3(-1)(-5) + 0 + (-1)(4)$$

$$= 11$$

5.8 THE COFACTOR MATRIX

Associated with a matrix **A** is another matrix whose components are co-factors of the corresponding components of **A**. Such a matrix is called the *cofactor matrix*. Consider, for example, the matrix

$$\mathbf{A} = \begin{bmatrix} a_{11} & a_{12} & a_{13} \\ a_{21} & a_{22} & a_{23} \\ a_{31} & a_{32} & a_{33} \end{bmatrix}$$

The corresponding cofactor matrix is

$$\mathbf{A}_{\text{cofactor}} = \begin{bmatrix} C_{11} & C_{12} & C_{13} \\ C_{21} & C_{22} & C_{23} \\ C_{31} & C_{32} & C_{33} \end{bmatrix}$$

5.9 THE ADJOINT MATRIX

The transpose of the cofactor matrix is called the *adjoint matrix*. Thus, for the above matrix **A**,

$$\mathbf{A}_{\text{adj}} = [\mathbf{A}_{\text{cofactor}}]^T$$

or

$$\mathbf{A}_{\text{adj}} = \begin{bmatrix} C_{11} & C_{12} & C_{13} \\ C_{21} & C_{22} & C_{23} \\ C_{31} & C_{32} & C_{33} \end{bmatrix}^T = \begin{bmatrix} C_{11} & C_{21} & C_{31} \\ C_{12} & C_{22} & C_{32} \\ C_{13} & C_{23} & C_{33} \end{bmatrix}$$

As we shall see in the next section, the adjoint matrix is very useful in finding the inverse of a given matrix.

5.10 THE INVERSE MATRIX

If, for a given square matrix **A**, there exists another square matrix **B** such that **AB** = **BA** = **I** (the identity matrix), then **B** is said to be the *inverse* of **A**. The inverse of **A** is usually denoted as $\mathbf{A}^{-1}$. When it exists, $\mathbf{A}^{-1}$ plays a role similar to that played by the reciprocal of a given number in ordinary algebra, although it must be noted that not all matrices have inverses. As a matter of fact, only square matrices with nonzero determinants have inverses.

EXAMPLE 5.11

Let

$$A = \begin{bmatrix} 4 & 0 & 0 \\ 0 & 6 & 2 \\ 2 & 0 & 1 \end{bmatrix} \quad \text{and} \quad B = \begin{bmatrix} \frac{1}{4} & 0 & 0 \\ \frac{1}{6} & \frac{1}{6} & -\frac{1}{3} \\ -\frac{1}{2} & 0 & 1 \end{bmatrix}$$

Then

$$AB = \begin{bmatrix} 4 & 0 & 0 \\ 0 & 6 & 2 \\ 2 & 0 & 1 \end{bmatrix} \begin{bmatrix} \frac{1}{4} & 0 & 0 \\ \frac{1}{6} & \frac{1}{6} & -\frac{1}{3} \\ -\frac{1}{2} & 0 & 1 \end{bmatrix} = \begin{bmatrix} 1 & 0 & 0 \\ 0 & 1 & 0 \\ 0 & 0 & 1 \end{bmatrix}$$

$$BA = \begin{bmatrix} \frac{1}{4} & 0 & 0 \\ \frac{1}{6} & \frac{1}{6} & -\frac{1}{3} \\ -\frac{1}{2} & 0 & 1 \end{bmatrix} \begin{bmatrix} 4 & 0 & 0 \\ 0 & 6 & 2 \\ 2 & 0 & 1 \end{bmatrix} = \begin{bmatrix} 1 & 0 & 0 \\ 0 & 1 & 0 \\ 0 & 0 & 1 \end{bmatrix}$$

That is,

$$AB = BA = I$$

Hence, by definition, $B = A^{-1}$. It may be noted that if a given matrix has an inverse, the inverse is unique.

5.11 HOW TO FIND AN INVERSE

We shall illustrate two methods for finding the inverse (if it exists) of a square matrix.

5.11.1 The Direct Method

This method is quite easy to grasp, because it is based on the definition of the inverse.

EXAMPLE 5.12

Find the inverse of

$$A = \begin{bmatrix} 4 & 2 \\ 1 & 3 \end{bmatrix}$$

We know that the inverse of A is another matrix, A^{-1}, such that $AA^{-1} = I$. Let

$$A^{-1} = \begin{bmatrix} b_{11} & b_{12} \\ b_{21} & b_{22} \end{bmatrix}$$

Then, by definition,

$$\mathbf{A}\mathbf{A}^{-1} = \begin{bmatrix} 4 & 2 \\ 1 & 3 \end{bmatrix} \begin{bmatrix} b_{11} & b_{12} \\ b_{21} & b_{22} \end{bmatrix} = \begin{bmatrix} 1 & 0 \\ 0 & 1 \end{bmatrix}$$

From the definition of equality of matrices, we get

$$4b_{11} + 2b_{21} = 1 \tag{5.2}$$
$$4b_{12} + 2b_{22} = 0 \tag{5.3}$$
$$1b_{11} + 3b_{21} = 0 \tag{5.4}$$
$$1b_{12} + 3b_{22} = 1 \tag{5.5}$$

From Equation (5.3), $b_{12} = -\frac{1}{2}b_{22}$. When substituted in Equation (5.5), this gives $-\frac{1}{2}b_{22} + 3b_{22} = 1$ or $b_{22} = \frac{2}{5}$. Therefore,

$$b_{12} = -\frac{1}{2}b_{22} = -\frac{1}{2}(\frac{2}{5}) = -\frac{1}{5}$$

From Equation (5.4), $b_{11} = -3b_{21}$. When substituted in Equation (5.2), this gives

$$4(-3b_{21}) + 2b_{21} = 1$$

or $b_{21} = -\frac{1}{10}$. Therefore,

$$b_{11} = -3b_{21} = -3(-\frac{1}{10}) = \frac{3}{10}$$

Hence

$$\mathbf{A}^{-1} = \begin{bmatrix} b_{11} & b_{12} \\ b_{21} & b_{22} \end{bmatrix} = \begin{bmatrix} \frac{3}{10} & -\frac{1}{5} \\ -\frac{1}{10} & \frac{2}{5} \end{bmatrix}$$

Check:

$$\mathbf{A}\mathbf{A}^{-1} = \begin{bmatrix} 4 & 2 \\ 1 & 3 \end{bmatrix} \begin{bmatrix} \frac{3}{10} & -\frac{1}{5} \\ -\frac{1}{10} & \frac{2}{5} \end{bmatrix} = \begin{bmatrix} 1 & 0 \\ 0 & 1 \end{bmatrix} = \mathbf{I}$$

It is obvious that finding the inverse of a matrix of order higher than 2×2 would be a rather cumbersome task using the direct method.

5.11.2 Inverting a Matrix by Utilizing Its Determinant and Its Adjoint Matrix*

Without giving a formal proof, we define the inverse of a matrix $\mathbf{A}$ as

$$\mathbf{A}^{-1} = \frac{\mathbf{A}_{adj}}{|\mathbf{A}|}$$

* Another method of obtaining the inverse of a matrix is by utilizing *row operations*. The concept of row operations is described in Appendix E. Assume that we start with two matrices; matrix $\mathbf{A}$ and an identity matrix $\mathbf{I}$ of the same order as the matrix $\mathbf{A}$. We then perform row operations in such a way that matrix $\mathbf{A}$ is transformed to an identity matrix. If, for each row operation performed on $\mathbf{A}$, the same row operation is performed on $\mathbf{I}$, then by the time $\mathbf{A}$ is transformed to an identity matrix, the original matrix $\mathbf{I}$ would be transformed to the inverse of $\mathbf{A}$.

where the determinant $|\mathbf{A}| \neq 0$. We have already defined $|\mathbf{A}|$ and $\mathbf{A}_{adj}$ in Sections 5.7 and 5.9, respectively.

EXAMPLE 5.13

Find the inverse of

$$\mathbf{A} = \begin{bmatrix} 4 & 2 \\ 1 & 3 \end{bmatrix}$$

The determinant of the matrix $\mathbf{A}$ is

$$|\mathbf{A}| = 12 - 2 = 10$$

and the adjoint of matrix $\mathbf{A}$ is

$$\mathbf{A}_{adj} = \begin{bmatrix} C_{11} & C_{21} \\ C_{12} & C_{22} \end{bmatrix} = \begin{bmatrix} 3 & -2 \\ -1 & 4 \end{bmatrix}$$

Hence

$$\mathbf{A}^{-1} = \frac{\mathbf{A}_{adj}}{|\mathbf{A}|} = \frac{1}{10} \begin{bmatrix} 3 & -2 \\ -1 & 4 \end{bmatrix} = \begin{bmatrix} \frac{3}{10} & -\frac{1}{5} \\ -\frac{1}{10} & \frac{2}{5} \end{bmatrix}$$

Note that this result is the same as that obtained by the direct method.

EXAMPLE 5.14

Find the inverse of

$$\mathbf{A} = \begin{bmatrix} 4 & 0 & 0 \\ 0 & 6 & 2 \\ 2 & 0 & 1 \end{bmatrix}$$

Expanding along the first row, we find that

$$|\mathbf{A}| = a_{11}C_{11} + a_{12}C_{12} + a_{13}C_{13}$$
$$= 4(6) + 0 + 0 = 24$$

The adjoint matrix is

$$\mathbf{A}_{adj} = \begin{bmatrix} C_{11} & C_{21} & C_{31} \\ C_{12} & C_{22} & C_{32} \\ C_{13} & C_{23} & C_{33} \end{bmatrix} = \begin{bmatrix} 6 & 0 & 0 \\ 4 & 4 & -8 \\ -12 & 0 & 24 \end{bmatrix}$$

Now

$$A^{-1} = \frac{A_{adj}}{|A|} = \frac{1}{24} \begin{bmatrix} 6 & 0 & 0 \\ 4 & 4 & -8 \\ -12 & 0 & 24 \end{bmatrix}$$

$$= \begin{bmatrix} \frac{1}{4} & 0 & 0 \\ \frac{1}{6} & \frac{1}{6} & -\frac{1}{3} \\ -\frac{1}{2} & 0 & 1 \end{bmatrix}$$

The reader can check that $AA^{-1} = I$.

5.12 LINEAR INDEPENDENCE

A set of vectors $V_1, V_2, \ldots, V_m$ of the same dimension is said to be linearly *dependent* if a set of scalars $k_1, k_2, \ldots, k_m$, not all of which are zero, can be found such that

$$k_1 V_1 + k_2 V_2 + \cdots + k_m V_m = 0 \tag{5.6}$$

where 0 represents a zero vector.

If the above relationship holds *only* when all the scalars $k_1, k_2, \ldots, k_m$ are zero, the set of vectors $V_1, V_2, \ldots, V_m$ is said to be *linearly independent*. We can test linear independence by two methods: (1) by applying the definition of linear independence, and (2) by the use of determinants.

5.12.1 Testing Linear Independence by Using Its Definition

EXAMPLE 5.15

Test the vectors

$$V_1 = \begin{bmatrix} 3 \\ 2 \end{bmatrix} \quad \text{and} \quad V_2 = \begin{bmatrix} 6 \\ 4 \end{bmatrix}$$

for linear independence. Let

$$k_1 V_1 + k_2 V_2 = 0$$

or

$$k_1 \begin{bmatrix} 3 \\ 2 \end{bmatrix} + k_2 \begin{bmatrix} 6 \\ 4 \end{bmatrix} = \begin{bmatrix} 0 \\ 0 \end{bmatrix}$$

or

$$3k_1 + 6k_2 = 0 \tag{5.7}$$

$$2k_1 + 4k_2 = 0 \tag{5.8}$$

Clearly, Equations (5.7) and (5.8) are simultaneously satisfied if we let $k_1 = 1$ and $k_2 = -\frac{1}{2}$. Hence, according to the definition given in Equation (5.6), vectors $\mathbf{V}_1$ and $\mathbf{V}_2$ are linearly dependent. That is, $\mathbf{V}_1$ and $\mathbf{V}_2$ do *not* constitute a linearly independent set.

EXAMPLE 5.16

Test the vectors

$$\mathbf{V}_3 = \begin{bmatrix} 2 \\ 3 \end{bmatrix} \quad \text{and} \quad \mathbf{V}_4 = \begin{bmatrix} 4 \\ 2 \end{bmatrix}$$

for linear independence. Let

$$k_3 \mathbf{V}_3 + k_4 \mathbf{V}_4 = 0$$

or

$$k_3 \begin{bmatrix} 2 \\ 3 \end{bmatrix} + k_4 \begin{bmatrix} 4 \\ 2 \end{bmatrix} = \begin{bmatrix} 0 \\ 0 \end{bmatrix}$$

or

$$2k_3 + 4k_4 = 0 \tag{5.9}$$

$$3k_3 + 2k_4 = 0 \tag{5.10}$$

Clearly, Equations (5.9) and (5.10) are simultaneously satisfied *only* if $k_3 = 0$ and $k_4 = 0$. Hence the vectors $\mathbf{V}_3$ and $\mathbf{V}_4$ constitute a linearly independent set.

5.12.2 Testing Linear Independence by the Use of Determinants

Calculate the numerical value of the determinant associated with the set of vectors. As mentioned previously, the determinant is defined only for square matrices. If the determinant is nonzero, the set of vectors is *linearly independent*; if it is zero, the set is *linearly dependent*.

EXAMPLE 5.17

We test the vector set of Example 5.15.

$$\mathbf{V}_1 = \begin{bmatrix} 3 \\ 2 \end{bmatrix} \quad \text{and} \quad \mathbf{V}_2 = \begin{bmatrix} 6 \\ 4 \end{bmatrix}$$

The determinant of the equivalent matrix is $\begin{vmatrix} 3 & 6 \\ 2 & 4 \end{vmatrix}$, and its numerical value is $(3)(4) - (2)(6) = 0$. Therefore, the set $\mathbf{V}_1, \mathbf{V}_2$ is a linearly dependent set.

EXAMPLE 5.18

We test the vector set of Example 5.16.

$$\mathbf{V}_3 = \begin{bmatrix} 2 \\ 3 \end{bmatrix} \quad \text{and} \quad \mathbf{V}_4 = \begin{bmatrix} 4 \\ 2 \end{bmatrix}$$

The determinant of the equivalent matrix is $\begin{vmatrix} 2 & 4 \\ 3 & 2 \end{vmatrix}$, and its numerical value is $(2)(2) - (3)(4) = -8$. Since the value of the determinant is *nonzero*, the set $\mathbf{V}_3$, $\mathbf{V}_4$ is a linearly independent set.

5.13 MATRIX AND VECTOR REPRESENTATION OF LINEAR SYSTEMS, AND VICE VERSA

Consider a system of two linear equations

$$2x_1 + 4x_2 = b_1 \tag{5.11}$$

$$3x_1 + 2x_2 = b_2 \tag{5.12}$$

If we use matrix notation, this system can be written as

$$\underset{\Downarrow}{\mathbf{A}} \quad \times \quad \underset{\Downarrow}{\mathbf{X}} \quad = \quad \underset{\Downarrow}{\mathbf{b}}$$

$$\begin{bmatrix} 2 & 4 \\ 3 & 2 \end{bmatrix} \begin{bmatrix} x_1 \\ x_2 \end{bmatrix} = \begin{bmatrix} b_1 \\ b_2 \end{bmatrix} \tag{5.13}$$

Equation (5.13) is equivalent to (5.11) and (5.12), as we can verify by matrix multiplication of the left-hand side of Equation (5.13) and equating it to its right-hand side.

A system of the type given in Equation (5.13) reflects the structural linear constraints of a typical linear programming problem. The matrix $\mathbf{A}$ is called the input–output *coefficient matrix*, $\mathbf{b}$ is the *constant vector* reflecting resource capacities or the requirements, and $\mathbf{X}$ is the *solution vector*.

The two linear equations (5.11) and (5.12) can also be represented with the aid of vectors as follows:

$$\begin{bmatrix} 2 \\ 3 \end{bmatrix} x_1 + \begin{bmatrix} 4 \\ 2 \end{bmatrix} x_2 = \begin{bmatrix} b_1 \\ b_2 \end{bmatrix} \tag{5.14}$$

It is obvious that Equation (5.14), which is a vector representation, can be translated back to the linear equations (5.11) and (5.12).

5.14 VECTOR SPACE

Consider the system in Equation (5.14). The problem is usually to find appropriate values of x_1 and x_2 for a given set of values b_1 and b_2. Now, suppose that we generate all possible values of b_1 and b_2 by varying the values of x_1 and x_2. Then the set of vectors $\begin{bmatrix} b_1 \\ b_2 \end{bmatrix}$ generated by all possible choices of x_1 and x_2 is called a *vector space* of two dimensions. Similarly, the set of *all* three-component vectors forms a three-dimensional vector space, and so on. The idea of the vector space is important in linear programming problems because the resource capacity, or the requirement vector, is located in a particular vector space. Our task is to choose a set of vectors that can *span* the particular vector space.

5.15 BASIS FOR A VECTOR SPACE

A *basis* for a vector space is a set of linearly independent vectors such that *any* vector in the vector space can be expressed as a linear combination of this set. This set spans the entire vector space. The vectors in such a set are called the *basic vectors*.

Let us consider two linearly independent vectors*

$$\mathbf{V}_1 = \begin{bmatrix} 1 \\ 0 \end{bmatrix} \quad \text{and} \quad \mathbf{V}_2 = \begin{bmatrix} 0 \\ 1 \end{bmatrix}$$

Any two-dimensional vector can be expressed as a linear combination of these two linearly independent vectors. For example,

$$\mathbf{V}_3 = \begin{bmatrix} 200 \\ 100 \end{bmatrix}$$

can be represented as $\mathbf{V}_3 = 200\mathbf{V}_1 + 100\mathbf{V}_2$. We come to the conclusion that vectors

$$\mathbf{V}_1 = \begin{bmatrix} 1 \\ 0 \end{bmatrix} \quad \text{and} \quad \mathbf{V}_2 = \begin{bmatrix} 0 \\ 1 \end{bmatrix}$$

* By applying the definition of linear independence, the reader can check that

$$\mathbf{V}_1 = \begin{bmatrix} 1 \\ 0 \end{bmatrix} \quad \text{and} \quad \mathbf{V}_2 = \begin{bmatrix} 0 \\ 1 \end{bmatrix}$$

are linearly independent.

are one set of basic vectors for a two-dimensional vector space. The basis is

$$[V_1 \quad V_2] = \begin{bmatrix} 1 & 0 \\ 0 & 1 \end{bmatrix}$$

Similarly, in a three-dimensional vector space, we need a set of three linearly independent vectors to form the basis. The argument can easily be extended to n dimensions.

It should be noted that linear independence is a necessary condition for forming a basis for a vector space in order that any vector in that space may be representable as a linear combination of the basic vectors. To emphasize this point, let us again consider vector

$$V_3 = \begin{bmatrix} 200 \\ 100 \end{bmatrix}$$

Clearly, $V_3 = 200V_1 + 100V_2$, provided that

$$V_1 = \begin{bmatrix} 1 \\ 0 \end{bmatrix} \quad \text{and} \quad V_2 = \begin{bmatrix} 0 \\ 1 \end{bmatrix}$$

where V_1 and V_2 are linearly independent. But if we consider two linearly dependent vectors, say

$$V_4 = \begin{bmatrix} 3 \\ 2 \end{bmatrix} \quad \text{and} \quad V_5 = \begin{bmatrix} 6 \\ 4 \end{bmatrix}$$

then we simply cannot express V_3 as a linear combination of V_4 and V_5. The reader should verify this statement with a geometrical representation of these vectors.

The concepts of linear independence and basis are very important in linear programming. As we shall observe in Chapter 7, the first tableau of the simplex method creates a basis, say, for an m-dimensional space, by using m linearly independent unit vectors. Further, a "degeneracy" occurs in linear programming problems when an m-dimensional vector is represented as a linear combination of less than m independent vectors.*

5.16   SYSTEMS OF SIMULTANEOUS LINEAR EQUATIONS

Since the structural constraints of a linear programming problem can be expressed as a system of simultaneous linear equations, a knowledge of such a system is important. Finding a solution to a system of linear equations requires the assignment of values to $x_1, x_2, \ldots$, so that *all* the equations are satisfied. When such a system is expressed in the matrix notation as $AX = b$,

---

* See Sections 3.5.2 and 7.7.

its solution involves the determination of the components of the vector $\mathbf{X}$. A system may have a unique solution, no solution, or an infinite number of solutions.

We classify a system of simultaneous linear equations into three categories.

CATEGORY 1

The system of linear equations contains n equations and n unknowns. In this category, we can have one of three cases:
1. The system has a unique solution (see Appendix B).
2. The system is inconsistent and has no solution (see Appendix C).
3. The system has an infinite number of solutions (see Appendix D).

CATEGORY 2

The system of linear equations has more equations than unknowns. This type of system of linear equations results whenever there is either redundancy (i.e., one or more equations in the system can be obtained by forming linear combinations of the remaining equations) or inconsistency. When the redundancy exists, the system can be reduced, by eliminating the redundancy, to that of category 1.

CATEGORY 3

The system of linear equations has more unknowns than equations. In this case, the system has no solution if the equations are inconsistent. However, if the system can be solved, it has an infinite number of solutions. Various possible solutions are identified by assigning an arbitrary value(s) to the excess of unknowns over equations. Of these, one or more are chosen in order to obtain an optimal value of some objective function.

The reader will realize that linear programming problems fall into category 3.

5.17 TERMINOLOGY OF LINEAR PROGRAMMING SOLUTIONS

Consider now the general linear model of a linear programming problem:

Maximize* $\qquad p_1 x_1 + p_2 x_2 + \cdots + p_n x_n$

subject to the linear constraints†

$$a_{11}x_1 + a_{12}x_2 + \cdots + a_{1n}x_n \leqslant b_1$$
$$a_{21}x_1 + a_{22}x_2 + \cdots + a_{2n}x_n \leqslant b_2$$
$$\vdots \qquad \vdots \qquad \qquad \vdots \qquad \vdots$$
$$a_{m1}x_1 + a_{m2}x_2 + \cdots + a_{mn}x_n \leqslant b_m$$

---

* This corresponds to the objective function involving profit contribution. See Section 1.2.
† This corresponds to the technical specifications of Table 1.1.

and subject to the nonnegativity constraints*

$$x_1 \geqslant 0 \quad x_2 \geqslant 0 \cdots x_n \geqslant 0$$

The linear constraints of the "less than or equal to" type are transformed into equations by the addition of nonnegative slack variables. When the above problem is stated in the form of equations, we have $m + n$ unknown variables (n real variables and m slack variables). Thus, when stated in the form of equations, a linear programming problem is such that the number of equations is less than the number of unknowns. Hence an infinite number of solutions can be found. Of these, the solution that optimizes the objective function is chosen.

Obtaining a solution to a linear programming problem involves assigning specific values to the unknown real variables $(x_1, x_2, \ldots, x_n)$ and the unknown slack or surplus variables $(x_{n+1}, x_{n+2}, \ldots, x_{n+m})$ without violating the given structural constraints and the nonnegativity constraints. The following terminology, depending upon whether or not the solution is an extreme-point solution and depending upon the number of positive components in the solution vector, is used to identify different types of solutions:†

1. *Nonextreme-point solution*; any solution that does not lie at the corner point of the convex set and contains more than m positive components.
2. *Extreme-point solution*; any solution that lies at a corner point of the convex set. There are two types of extreme-point solutions (1) basic feasible, and (2) degenerate basic feasible.
   a. *Basic feasible solution*; any solution that contains exactly m positive components.
   b. *Degenerate basic feasible solution*; any solution that contains less than m positive components.
3. *Optimal solution*; any basic feasible or degenerate basic feasible solution that either maximizes or minimizes the objective function.

A schematic representation of the above classification is shown in Figure 5.7.

5.18 SUMMARY

The purpose of this chapter was to familiarize the reader with some basic concepts of matrix algebra so that linear systems can be expressed in matrix and vector notation. This should enable us to understand more fully the rationale and the mechanics of the simplex method.

---

* These are the usual nonnegativity constraints of a linear programming problem.

† Where m equals the number of linear-structural constraints.

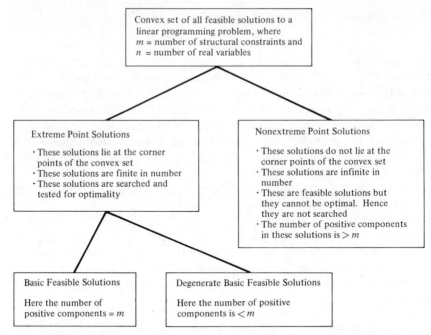

Figure 5.7

We defined such concepts as matrices, vectors, determinants, vector space, and a basis for a vector space. Addition, subtraction, and multiplication operations involving matrices and vectors were briefly illustrated. The concept of an inverse matrix was discussed and a general method for finding the inverse of a square matrix was illustrated. We then described three different categories of systems of linear equations and explained how a system of linear equations can have a unique solution, no solution, or an infinite number of solutions. A linear programming problem falls under that category in which the number of unknowns is more than the number of equations. Hence a linear programming problem has an infinite number of solutions. However, the search for the optimal solution is conducted by testing only the extreme-point solutions.

Finally, we summarized the terminology that can be used to distinguish different types of linear programming solutions, depending upon the relationship between the number of positive components in the solution and the number of structural constraints m. In particular, we defined these types of linear programming solutions: (1) nonextreme point, (2) basic feasible, (3) degenerate basic feasible, and (4) optimal.

REFERENCES

Bellman, R. *Introduction to Matrix Analysis*, 2nd ed. New York: McGraw-Hill Book Company, 1970.

Childress, R. L. *Sets, Matrices, and Linear Programming*. Englewood Cliffs, N.J.: Prentice-Hall, Inc., 1974.

Loomba, N. P., and E. Turban. *Applied Programming for Management*. New York: Holt, Rinehart and Winston, Inc., 1974.

Reiner, I. *Introduction to Matrix Theory and Linear Algebra*. New York: Holt, Rinehart and Winston, Inc., 1971.

Schneider, H., and Barker, G. P. *Matrices and Linear Algebra*, 2nd ed. New York: Holt, Rinehart and Winston, Inc., 1973.

Shank, J. K. *Matrix Methods in Accounting: An Introduction*. Reading, Mass.: Addison-Wesley Publishing Co., Inc., 1972.

REVIEW QUESTIONS AND PROBLEMS

5.1. How does a determinant differ from a matrix?

5.2. Consider the 3 × 3 matrix.

$$A = \begin{bmatrix} 2 & 0 & -1 \\ 1 & 2 & 4 \\ 0 & 3 & 2 \end{bmatrix}$$

Express **A** as a set of row vectors, and as a set of column vectors. Evaluate the determinant of **A**.

5.3. Consider the following matrices:

$$A = \begin{bmatrix} 3 & 2 & 0 \\ 1 & 2 & 3 \\ 4 & 0 & 1 \end{bmatrix} \qquad B = \begin{bmatrix} 2 & 4 \\ 3 & 0 \\ 1 & 2 \end{bmatrix}$$

$$X = \begin{bmatrix} 1 & 4 \\ 1 & 2 \\ 4 & 0 \end{bmatrix} \qquad Y = \begin{bmatrix} 2 & 4 & 0 \\ 1 & -4 & 2 \end{bmatrix}$$

$$R = \begin{bmatrix} 1 & 4 & 0 \\ 1 & 2 & 3 \\ -6 & 1 & 2 \end{bmatrix} \qquad S = \begin{bmatrix} 1 & 0 & 0 \\ 0 & 1 & 0 \\ 0 & 0 & 1 \end{bmatrix}$$

Find **AB**, **BA**, **XY**, **YX**, **RS**, **SR**, **ABY**, S^2, and $(A - R)S$.

5.4. Consider the following matrices:

$$A = \begin{bmatrix} 1 & 2 & 0 \\ 0 & 1 & 3 \\ 0 & 2 & 8 \end{bmatrix} \qquad B = \begin{bmatrix} 1 & -8 & 3 \\ 0 & 4 & -\frac{3}{2} \\ 0 & -1 & \frac{1}{2} \end{bmatrix}$$

Find **AB** and **BA**. What is the inverse of **A**?

5.5. Consider the following matrices:

$$A = \begin{bmatrix} 3 & 0 & 1 \\ 2 & 4 & 3 \end{bmatrix} \qquad B = \begin{bmatrix} 3 & 7 & 4 \\ 5 & 3 & 3 \\ 1 & 1 & 2 \end{bmatrix}$$

$$C = \begin{bmatrix} 4 & 3 \\ 1 & 0 \\ 6 & 2 \end{bmatrix} \qquad D = \begin{bmatrix} 3 & 4 & 2 \\ 6 & 1 & 0 \end{bmatrix}$$

$$V_1 = \begin{bmatrix} 2 \\ 4 \\ 5 \end{bmatrix} \qquad V_2 = \begin{bmatrix} 5 & 0 & 3 \end{bmatrix}$$

Find the following:
(a) $A + B$, $A + D$, $C - D$, CA, CB, BV_1, BV_2, V_1B, and V_1A.
(b) k_1A and Ak_1, where $k_1 = 2$.
(c) A^T, B^T, and V_1^T.

5.6. (a) Explain the meaning of linear independence, vector space, and basis for a vector space.
(b) Express

$$P_0 = \begin{bmatrix} 10 \\ 4 \\ 6 \end{bmatrix}$$

as a linear combination of P_1, P_2, and P_3, where

$$P_1 = \begin{bmatrix} 1 \\ 0 \\ 0 \end{bmatrix} \quad P_2 = \begin{bmatrix} 4 \\ 1 \\ 2 \end{bmatrix} \quad P_3 = \begin{bmatrix} 2 \\ -1 \\ 4 \end{bmatrix} \quad P_4 = \begin{bmatrix} 2 \\ 4 \\ 10 \end{bmatrix}$$

What can you say about linear independence among P_1, P_2, P_3, and P_4? Discuss your answer.

5.7. (a) Distinguish between a minor and a cofactor of a matrix; between the cofactor matrix and the adjoint matrix.
(b) Consider the following matrix:

$$A = \begin{bmatrix} 2 & 0 & -1 \\ 1 & 2 & 4 \\ 0 & 3 & 2 \end{bmatrix}$$

(1) Find $|A|$.
(2) Find A_{adj} and $A_{cofactor}$.
(3) Find A^{-1}.

5.8. Find solutions to the following systems of equations:

(a)
$$\begin{aligned} 5X_1 + 4X_2 + X_3 &= 100 \\ 2X_2 + 4X_3 &= 80 \\ 2X_1 + 6X_2 &= 90 \end{aligned}$$

(b)
$$\begin{aligned} 5X_1 + 15X_2 &= 100 \\ 2X_2 + 4X_3 &= 80 \\ 2X_1 + 6X_2 &= 90 \end{aligned}$$

(c) Write the coefficient matrix of Problem 5.8(a) and find its determinant.

6
The Vector Method

6.1 INTRODUCTION

It was illustrated in Section 5.13 how a set of simultaneous linear equations can be represented as vectors. Since the structural constraints of a linear programming problem can be expressed as linear equations, we can represent them by employing vector notation. The solution of the linear programming problem, then, can be obtained by performing certain vector operations. This approach to solving linear programming problems will be referred to as the *vector method*.

The linear programming problem of Table 3.2 has now been solved by graphical method (Chapter 3) and the algebraic method (Chapter 4). In this chapter, the same problem will be solved by the vector method. A knowledge of the vector method will help the reader to understand the mechanics and rationale of the simplex method to be presented in Chapter 7.

For quick reference, the data of Table 3.2 are listed in Table 6.1.

Table 6.1

Department	Product		Capacity per Time Period
	A	B	
Cutting	10	6	2,500
Folding	5	10	2,000
Packaging	1	2	500
Profit per Unit	$23	$32	

6.2 RATIONALE FOR THE VECTOR METHOD

Before illustrating the mechanics of the vector method, we briefly describe its rationale. It was explained in Section 5.6.6 that a linear programming problem boils down to expressing an m-dimensional vector (the capacity or requirement vector whose components are the right-hand-side constants of the structural constraints) as a linear combination of a set of linearly independent vectors. A total of $m + n$ vectors are available (m vectors corresponding to slack or surplus variables and n vectors corresponding to real variables), out of which a set of m linearly independent vectors is chosen.* Once such a basic set is identified (i.e., a program is designed) it divides the total available $m + n$ vectors into a set of m basic vectors and n nonbasic vectors. The next step is to test the optimality of the program by, first, expressing *each* nonbasic vector as a linear combination of the current basic vectors and, then, calculating the *net* effect of such an exchange or replacement. If any of the net effect is positive (for maximization case) or negative (for minimization case) the program is not optimal. A new program is then designed by bringing in *one* of the nonbasic variables, and the new program is tested for optimality. The iterative process is repeated until an optimal program has been designed.

In the next section we illustrate the vector method by applying it to the problem of Table 6.1.

6.3 VECTOR REPRESENTATION OF THE PROBLEM

By use of the arguments of the algebraic method (Section 4.4.2), we can express the data of Table 6.1 in the form of equations:

$$10X + 6Y + 1S_1 + 0S_2 + 0S_3 = 2{,}500 \qquad (6.1)$$

$$5X + 10Y + 0S_1 + 1S_2 + 0S_3 = 2{,}000 \qquad (6.2)$$

$$1X + 2Y + 0S_1 + 0S_2 + 1S_3 = 500 \qquad (6.3)$$

Equations (6.1) to (6.3) are the structural constraints. The objective function is

$$23X + 32Y + 0(S_1 + S_2 + S_3) \qquad (6.4)$$

The problem is to maximize the objective function subject to linear constraints (6.1) to (6.3) and the nonnegativity constraints $X \geqslant 0$, $Y \geqslant 0$, $S_1 \geqslant 0$, $S_2 \geqslant 0$, and $S_3 \geqslant 0$. Writing Equations (6.1) to (6.3) in the vector form, we obtain

$$X \begin{bmatrix} 10 \\ 5 \\ 1 \end{bmatrix} + Y \begin{bmatrix} 6 \\ 10 \\ 2 \end{bmatrix} + S_1 \begin{bmatrix} 1 \\ 0 \\ 0 \end{bmatrix} + S_2 \begin{bmatrix} 0 \\ 1 \\ 0 \end{bmatrix} + S_3 \begin{bmatrix} 0 \\ 0 \\ 1 \end{bmatrix} = \begin{bmatrix} 2{,}500 \\ 2{,}000 \\ 500 \end{bmatrix}$$

* See the corresponding analysis in the algebraic method in Section 4.3.

If we let

$$P_1 = \begin{bmatrix} 10 \\ 5 \\ 1 \end{bmatrix} \quad P_2 = \begin{bmatrix} 6 \\ 10 \\ 2 \end{bmatrix} \quad P_3 = \begin{bmatrix} 1 \\ 0 \\ 0 \end{bmatrix}$$

$$P_4 = \begin{bmatrix} 0 \\ 1 \\ 0 \end{bmatrix} \quad P_5 = \begin{bmatrix} 0 \\ 0 \\ 1 \end{bmatrix} \quad P_0 = \begin{bmatrix} 2,500 \\ 2,000 \\ 500 \end{bmatrix}$$

then

$$XP_1 + YP_2 + S_1P_3 + S_2P_4 + S_3P_5 = P_0 \qquad (6.5)$$

Here P_1 and P_2 are the structural vectors; P_3, P_4, and P_5 are the unit vectors; P_0 is the constant or requirement vector. Equation (6.5) states the problem in simple terms. P_0 is a three-component vector that must be expressed as a linear combination of P_1, P_2, P_3, P_4, and P_5. In other words, the scalars X, Y, S_1, S_2, and S_3 are to be given nonnegative values such that Equation (6.5) is satisfied. As can be ascertained quickly, and as we noted while solving the same problem by the graphical method, an infinite number of values of the scalars can be found to satisfy Equation (6.5). Our objective is to choose that set of values which maximizes the profit function given by Equation (6.4).

One additional point must be emphasized. Since P_0 is a three-dimensional vector, we do not need more than three linearly independent vectors to represent it in a unique fashion. Accordingly, the problem can be solved as follows:

1. Express P_0 in terms of all possible linear combinations of P_1, P_2, P_3, P_4, and P_5 taken three at a time.
2. Calculate the profit contribution that results from each such combination, and choose that combination which yields the highest profit.

Theoretically, then, we shall have to test as many as 10 combinations in this simple problem.* This, however, would involve rather lengthy calculations. Instead, to save time and effort and to evolve a systematic method of search, we should like to be able to choose a particular combination of three linearly independent vectors as a starting point, and then progressively improve our solution. The method of improvement should be

* By definition, the combination of n things taken r at a time is given by

$$\binom{n}{r} = \frac{n!}{r!(n-r)!}$$

In this case, we have

$$\binom{5}{3} = \frac{5!}{3!2!} = 10$$

such that combinations giving higher profits than the current program could be immediately identified. In the vector method, this is accomplished in the following manner:

First, an initial program is designed in such a manner that it represents a basic feasible solution.* Second, to determine whether the initial or current program can be improved, the net effect on the objective function of introducing one of the nonbasic vectors to replace at least one of the basic vectors is tested.† If the objective function can be improved, this replacement is made. In other words, the old basis is replaced by a new basis. This testing for optimality and the replacement process is continued until the optimal set of basic vectors expressing P_0 is determined.

Proceeding in the above fashion (rather than checking each possible combination of the given vectors) reduces the computational work considerably. This, then, is the essence of the vector method.

6.4   ILLUSTRATION OF THE VECTOR METHOD

6.4.1   Design an Initial Program

As a first step, let us express the requirement vector P_0 as a linear combination of the set of vectors P_3, P_4, and P_5.‡ Our choice of vectors P_3, P_4, and P_5 as the basic vectors means that we are letting the scalars X and Y equal zero. In terms of the graphical method, this is equivalent to starting the initial solution at the origin of the three-dimensional space. Physically speaking, this corresponds to producing nothing and thereby letting all the resource capacities stay idle.

---

* This means, as the reader will recall from Section 5.17, that there are exactly three positive components (since this is a three-dimensional problem) in the solution. In vector terminology, this means that a linear combination of three vectors is used to represent the requirement vector P_0. Further, the three vectors in the linear combination are the basic vectors; the remaining two are the nonbasic vectors. The set of vectors in the solution constitutes a basis (see Section 5.15).

† This test is made for each of the nonbasic vectors. Then that nonbasic vector which shows the highest per unit improvement potential is exchanged with one (or more) of the current basic vectors in order to arrive at a new basis.

‡ As we observed in Chapter 5,

$$P_3 = \begin{bmatrix} 1 \\ 0 \\ 0 \end{bmatrix} \quad P_4 = \begin{bmatrix} 0 \\ 1 \\ 0 \end{bmatrix} \quad P_5 = \begin{bmatrix} 0 \\ 0 \\ 1 \end{bmatrix}$$

are linearly independent and thus comprise a set of basic vectors for a three-dimensional space. Hence the initial basis $\mathbf{B}$ is an identity matrix:

$$\mathbf{B} = [P_3\,P_4\,P_5] = \begin{bmatrix} 1 & 0 & 0 \\ 0 & 1 & 0 \\ 0 & 0 & 1 \end{bmatrix}$$

In Equation (6.5), if we let X and Y equal zero, we obtain

$$0\mathbf{P}_1 + 0\mathbf{P}_2 + S_1\mathbf{P}_3 + S_2\mathbf{P}_4 + S_3\mathbf{P}_5 = \mathbf{P}_0 \qquad (6.6)$$

or

$$S_1\begin{bmatrix}1\\0\\0\end{bmatrix} + S_2\begin{bmatrix}0\\1\\0\end{bmatrix} + S_3\begin{bmatrix}0\\0\\1\end{bmatrix} = \begin{bmatrix}2{,}500\\2{,}000\\500\end{bmatrix}$$

Obviously, if we let $S_1 = 2{,}500$, $S_2 = 2{,}000$, and $S_3 = 500$, the above equation is satisfied. Our initial solution, therefore, is $X = 0$, $Y = 0$, $S_1 = 2{,}500$, $S_2 = 2{,}000$, and $S_3 = 500$. Hence our first or initial program is

Program 1

$S_1 = 2{,}500$	$S_2 = 2{,}000$	$S_3 = 500$	$\Rightarrow \mathbf{P}_3, \mathbf{P}_4, \mathbf{P}_5$ basic vectors
$X = \quad 0$	$Y = \quad 0$		$\Rightarrow \mathbf{P}_1, \mathbf{P}_2$ nonbasic vectors

profit for program $1 = 0$

Substitution of the above values for X, Y, S_1, S_2, and S_3 into the objective function (6.4) shows that the profit for program 1 is zero.

In view of program 1, Equation (6.6) becomes

$$0\mathbf{P}_1 + 0\mathbf{P}_2 + 2{,}500\mathbf{P}_3 + 2{,}000\mathbf{P}_4 + 500\mathbf{P}_5 = \mathbf{P}_0 \qquad (6.7)$$

6.4.2   Test for Optimality (and Identify the Incoming Vector)

Is our initial program optimal? Since its profit contribution was found to be zero, the answer is obviously "no." However, we can formalize the optimality test by, first, expressing *each* nonbasic vector in terms of *current* basic vectors and then calculating the net effect of such an exchange or replacement. If *any* of the *net effect* is positive (for the maximization case) or negative (for the minimization case), the program is not optimal.

We first express the nonbasic vector $\mathbf{P}_1$ in terms of $\mathbf{P}_3$, $\mathbf{P}_4$, and $\mathbf{P}_5$. Let

$$a\mathbf{P}_3 + b\mathbf{P}_4 + c\mathbf{P}_5 = \mathbf{P}_1 \qquad (6.8)$$

Then

$$a\begin{bmatrix}1\\0\\0\end{bmatrix} + b\begin{bmatrix}0\\1\\0\end{bmatrix} + c\begin{bmatrix}0\\0\\1\end{bmatrix} = \begin{bmatrix}10\\5\\1\end{bmatrix}$$

The above equation is obviously satisfied if we let $a = 10$, $b = 5$, and $c = 1$. Thus Equation (6.8) becomes

$$10\mathbf{P}_3 + 5\mathbf{P}_4 + 1\mathbf{P}_5 = \mathbf{P}_1 \qquad (6.9)$$

The meaning of Equation (6.9) is that, at this solution stage, 1 unit of P_1 (having a profit of \$23 per unit) will replace 10 units of P_3, 5 units of P_4, and 1 unit of P_5 (P_3, P_4, P_5 each having a profit of zero per unit). Hence the net profit effect of introducing one unit of P_1 is

$$+23 - 10(0) - 5(0) - 1(0) = +23 \text{ dollars}$$

It can be similarly ascertained that our second nonbasic vector P_2 can be expressed as

$$6P_3 + 10P_4 + 2P_5 = P_2 \tag{6.10}$$

The net effect of introducing one unit of P_2 is

$$+32 - 6(0) - 10(0) - 2(0) = +32 \text{ dollars}$$

It is seen that the per unit net advantage of introducing P_1 and P_2, at this stage, are the same as the profit coefficients of X and Y in Equation (6.4). However, it should be emphasized that such exchanges (between a nonbasic vector and the set of basic vectors) are different from stage to stage and hence must be examined and compared at each stage of the solution process.

Since the per unit net effect is greater for P_2 than for P_1, we identify P_2 as the incoming vector.

6.4.3 Revise the Initial Program

Having concluded that our initial program (i.e., program 1) is not optimal, we must design a new and better program. That is accomplished by expressing P_0 as a linear combination of a new set of basic vectors.* The new set is formed by bringing in one of the nonbasic vectors to replace at least one of the basic vectors in the current solution. This replacement process, it must be emphasized, must be conducted such that *only one* nonbasic vector is introduced at a time. The incoming nonbasic vector will usually replace exactly one of the basic vectors (except in cases of degeneracy, where it can replace more than one). Since our problem is not degenerate, the basis here will always consist of three vectors.

We now proceed to explain the mechanics of the replacement process with respect to our example. It has already been established that P_2 is the incoming vector and that each unit of P_2 adds a net profit of \$32 at this stage. This exchange of P_2 (for P_3, P_4, and P_5) being profitable, we shall carry it to the limit. In other words, we shall keep on bringing units of P_2 into the solution until one of the vectors P_3, P_4, or P_5 is removed from the

---

* The fact that P_0 is expressed as a linear combination of a set of basic vectors means that we are satisfying the linear structural constraints.

solution. Let us say that at most h units of P_2 can be brought in. Then, from Equation (6.10), we get

$$6hP_3 + 10hP_4 + 2hP_5 = hP_2$$

or

$$6hP_3 + 10hP_4 + 2hP_5 - hP_2 = 0 \qquad (6.11)$$

Let us now subtract Equation (6.11) from (6.7). Then

$$0P_1 + hP_2 + (2,500 - 6h)P_3 + (2,000 - 10h)P_4 + (500 - 2h)P_5 = P_0 \qquad (6.12)$$

In Equation (6.12), P_0 has been expressed as a linear combination of P_2, P_3, P_4, and P_5. However, as stated previously, we need only three vectors in the solution. One of the basic vectors P_3, P_4, or P_5 should therefore be removed from the current basis. That is, the coefficients of either P_3 or P_4, or P_5 in Equation (6.12), should be made zero by assigning appropriate value to h. To reduce the coefficient of P_3 to zero, let

$$2,500 - 6h = 0 \qquad \text{or} \qquad h = 416\tfrac{2}{3}$$

To reduce the coefficient of P_4 to zero, let

$$2,000 - 10h = 0 \qquad \text{or} \qquad h = 200 \rightarrow$$

To reduce the coefficient of P_5 to zero, let

$$500 - 2h = 0 \qquad \text{or} \qquad h = 250$$

This means that the maximum number of units of P_2 that can be brought into the solution is 200. In other words, $h = 200$ is the controlling or limiting number and is the maximum allowable value that can be inserted into Equation (6.12) without making any term on the left-hand side negative.* Thus P_4 is the outgoing vector, and we show this fact by placing an arrow next to the minimum value of h (i.e., the maximum units of the incoming vector).

Letting $h = 200$ in Equation (6.12), we have

$$0P_1 + 200P_2 + 1,300P_3 + 0P_4 + 100P_5 = P_0 \qquad (6.13)$$

The second program, therefore, is

Program 2

$Y = 200$	$S_1 = 1,300$	$S_3 = 100$	$\Rightarrow P_2, P_3, P_5$ basic vectors
$X = 0$	$S_2 = 0$		$\Rightarrow P_1, P_4$ nonbasic vectors

profit for program 2 $= 32(200) + 0(1,300) + 0(100) = \$6,400$

* We cannot let any term in Equation (6.12) become negative, as this would violate the nonnegativity constraints.

6.4.4 Test for Optimality

Is the second program our optimal program? As stated previously, we can answer this question only by testing the net effect on the objective function of bringing in one of the nonbasic vectors $\mathbf{P}_1$ or $\mathbf{P}_4$ to replace one of the present basic vectors $\mathbf{P}_2$, $\mathbf{P}_3$, or $\mathbf{P}_5$. Representing $\mathbf{P}_1$ as a linear combination of the three basic vectors ($\mathbf{P}_2$, $\mathbf{P}_3$, and $\mathbf{P}_5$) now in the solution, we obtain $a\mathbf{P}_2 + b\mathbf{P}_3 + c\mathbf{P}_5 = \mathbf{P}_1$, where a, b, and c are scalars. Thus

$$a\begin{bmatrix} 6 \\ 10 \\ 2 \end{bmatrix} + b\begin{bmatrix} 1 \\ 0 \\ 0 \end{bmatrix} + c\begin{bmatrix} 0 \\ 0 \\ 1 \end{bmatrix} = \begin{bmatrix} 10 \\ 5 \\ 1 \end{bmatrix}$$

This vector equation can be translated into the following equations:

$$6a + 1b + 0c = 10$$
$$10a + 0b + 0c = 5$$
$$2a + 0b + 1c = 1$$

Solving this system of equations, we get

$$a = \tfrac{1}{2} \qquad b = 7 \qquad c = 0$$

or

$$\tfrac{1}{2}\mathbf{P}_2 + 7\mathbf{P}_3 + 0\mathbf{P}_5 = \mathbf{P}_1 \tag{6.14}$$

In other words, to bring in 1 unit of $\mathbf{P}_1$ we shall have to remove from the solution $\tfrac{1}{2}$ unit of $\mathbf{P}_2$, 7 units of $\mathbf{P}_3$, and 0 units of $\mathbf{P}_5$. This exchange will have the following net effect on the profit function:

$$+1(23) - \tfrac{1}{2}(32) - 7(0) - 0(0) = 7 \text{ dollars}$$

Since the net effect of this exchange is positive, it indicates that program 2 is not optimal.

6.4.5 Revise Program 2

Since program 2 is not optimal, we shall design a new program by replacing at least one of the basic vectors now in the solution, namely, $\mathbf{P}_2$, $\mathbf{P}_3$, or $\mathbf{P}_5$, with $\mathbf{P}_1$.*

---

* Actually, the relative exchange profitabilities of all the nonbasic vectors (here $\mathbf{P}_1$ and $\mathbf{P}_4$) should have been compared and the nonbasic vector with the highest net advantage chosen to make the replacement. But here, as soon as it was found that the nonbasic vector $\mathbf{P}_1$ had improvement potential, we decided to obtain a new basis to include $\mathbf{P}_1$. A complete set of comparisons among all the nonbasic vectors was not made because our intent is only to focus on the vector method as a forerunner of the simplex method, which does make *all* such comparisons at each stage of the solution. As it turns out, $\mathbf{P}_1$ is the correct choice. The reader should verify this statement.

Following the same argument as before, let us say that at most k units of $\mathbf{P}_1$ can be brought in without violating the nonnegativity constraints. Then, from Equation (6.14), we get

$$\tfrac{1}{2}k\mathbf{P}_2 + 7k\mathbf{P}_3 + (0)k\mathbf{P}_5 = k\mathbf{P}_1$$

or

$$\tfrac{1}{2}k\mathbf{P}_2 + 7k\mathbf{P}_3 + (0)k\mathbf{P}_5 - k\mathbf{P}_1 = 0 \qquad (6.15)$$

We subtract Equation (6.15) from (6.13):

$$k\mathbf{P}_1 + (200 - \tfrac{1}{2}k)\mathbf{P}_2 + (1{,}300 - 7k)\mathbf{P}_3 + (0)\mathbf{P}_4 + 100\mathbf{P}_5 = \mathbf{P}_0 \quad (6.16)$$

In Equation (6.16), $\mathbf{P}_0$ has been expressed as a linear combination of $\mathbf{P}_1$, $\mathbf{P}_2$, $\mathbf{P}_3$, and $\mathbf{P}_5$. However, we really need only three vectors in the solution, and therefore one of the basic vectors $\mathbf{P}_2$, $\mathbf{P}_3$, or $\mathbf{P}_5$ should be removed from the current basis. That is, the coefficients of either $\mathbf{P}_2$ or $\mathbf{P}_3$ or $\mathbf{P}_5$ in Equation (6.16) should be made zero by assigning an appropriate value to k. To reduce the coefficient of $\mathbf{P}_2$ to zero, let

$$200 - \tfrac{1}{2}k = 0 \qquad \text{or} \qquad k = 400$$

To reduce the coefficient of $\mathbf{P}_3$ to zero, let

$$1{,}300 - 7k = 0 \qquad \text{or} \qquad k = \tfrac{1{,}300}{7} = 185\tfrac{5}{7} \rightarrow$$

The coefficient of $\mathbf{P}_5$ cannot be reduced to zero because, as can be seen in Equation (6.16), the scalar k is not present in the coefficient of $\mathbf{P}_5$. Thus $k = \tfrac{1{,}300}{7}$ is the limiting case. That is, we cannot bring in more than $\tfrac{1{,}300}{7}$ units of $\mathbf{P}_1$ without violating the nonnegativity constraints. Substituting $k = \tfrac{1{,}300}{7}$ in Equation (6.16), we get

$$\tfrac{1{,}300}{7}\mathbf{P}_1 + \tfrac{750}{7}\mathbf{P}_2 + 0\mathbf{P}_3 + 0\mathbf{P}_4 + 100\mathbf{P}_5 = \mathbf{P}_0 \qquad (6.17)$$

The third program, therefore, is

Program 3

$X = \tfrac{1{,}300}{7}$	$Y = \tfrac{750}{7}$	$S_3 = 100$	$\Rightarrow \mathbf{P}_1, \mathbf{P}_2, \mathbf{P}_5$ basic vectors
$S_1 = 0$	$S_2 = 0$		$\Rightarrow \mathbf{P}_3, \mathbf{P}_4$ nonbasic vectors

profit for program 3 $= \tfrac{1{,}300}{7}(23) + \tfrac{750}{7}(32) = \$7{,}700$

6.4.6   Test for Optimality

As explained earlier, a given program can be tested for optimality by examining the net effect of bringing in each of the nonbasic variables. Therefore, we first express each nonbasic variable (here $\mathbf{P}_3$ and $\mathbf{P}_4$) in terms of current basic variables (here $\mathbf{P}_1$, $\mathbf{P}_2$, and $\mathbf{P}_5$) and then calculate their *net* effect on the profit function. If the above procedure is carried out for program 3, we shall find that program 3 is the optimal program, because, at this stage, the net effect of each nonbasic variable is negative.

We now summarize the two parts of the optimality test, which the reader can verify by proceeding in exactly the same manner as when we tested optimality of the first and the second programs:

Incoming Vector	Net Profit Change
$P_3 = -\frac{1}{7}P_1 + \frac{1}{14}P_2 - 0P_5$	$-\$1.00$
$P_4 = -\frac{3}{35}P_1 + \frac{1}{7}P_2 - \frac{1}{5}P_5$	$-\$2.60$

Since none of the nonbasic vectors will result in net change that is positive, the third program is the optimal program.

Various programs designed by the vector method correspond to the corner-point solutions of Figure 3.6, and we show this correspondence in Table 6.2. Also shown in Table 6.2 are the corresponding basic and nonbasic

Table 6.2

Program	Basic Variables	Basic Vectors	Nonbasic Variables	Nonbasic Vectors	Corresponding Solution from Figure 3.6	Total Profit
1	S_1, S_2, S_3	P_3, P_4, P_5	X, Y	P_1, P_2	O	\$0
2	Y, S_1, S_3	P_2, P_3, P_5	X, S_2	P_1, P_4	M	6,400
3	X, Y, S_3	P_1, P_2, P_5	S_1, S_2	P_3, P_4	H	7,700

variables of the various solutions derived by the algebraic method. Note that each specific program, regardless of the method used to design it, results in the production of the same products. All that we have done is this: We have analyzed the problem from different perspectives.

6.5 PROCEDURE SUMMARY FOR THE VECTOR METHOD (MAXIMIZATION CASE)

STEP 1. Formulate the problem.

a. Make a precise statement of the objective function and translate the technical specifications of the problem into inequalities. All variables are restricted to nonnegative values, and all inequalities are of the "less than or equal to" type.

b. Convert the inequalities into equalities by the addition of non-negative slack variables. Attach a per unit profit of zero to each of these slack variables or imaginary products.

c. Write the equations obtained in step b in the form of vectors.

STEP 2. Design a program (choose a set of basic vectors).

a. Express the requirement vector P_0 as a linear combination of a set of independent vectors (the first program is always obtained by expressing P_0 as a linear combination of unit vectors). This set of vectors forms the basis.

b. Express the program in the form of a vector equation that contains all basic and nonbasic vectors with proper scalars.

STEP 3. Test for optimality and identify the incoming vector.

Express each nonbasic vector as a linear combination of the basic vectors, and determine the exchange profitabilities of each nonbasic vector. If none of the nonbasic vectors shows any exchange profitability, the problem is solved and no revision is needed. Otherwise, the nonbasic vector with the highest exchange profitability is chosen as the incoming vector to become one of the new basic vectors.

STEP 4. Revise the program (choose a new set of basic vectors).

Determine the maximum units of the incoming vector. Multiply the exchange equation of the incoming vector by an unknown scalar k and arrange it in such a manner that the right-hand side is zero. Subtract this exchange equation from the program equation of step 2b. Then determine the maximum units of the incoming vector (the value of some constant k) so that at least one of the current basic vectors acquires zero as its scalar. This will lead to a new set of basic vectors and hence a new program, expressed in the form of a vector equation.

STEP 5. Obtain the optimal program.

Repeat steps 3 and 4 until an optimal program has been designed. An optimal program has been reached when the exchange profitabilities of all the nonbasic vectors are zero or negative.

A schematic of the vector method is shown in Figure 6.1.

6.6 SUMMARY

This chapter has presented both the rationale and the mechanics of the vector method. In Table 6.2 we have provided a comparison of the different solution stages (and their associated programs) encountered in solving the same problem by the graphical, algebraic, and vector methods. Such an analysis of the same problem from different perspectives gives us sufficient knowledge of linear programming, its structure, and its solution characteristics to better understand the material presented in the subsequent chapters of this book.

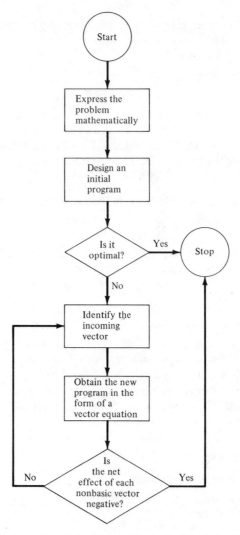

Figure 6.1   A Schematic of the Vector Method (Maximization Case)

REFERENCES

See references at the end of Chapter 3.

REVIEW QUESTIONS AND PROBLEMS

6.1. What is the rationale of the vector method?
6.2. Describe and compare the various steps of the vector and the algebraic methods.

6.3. Give the vector representation of the following linear programming problem:

Maximize $\qquad\qquad 5x + 6y$

subject to the constraints

$$4x + 7y \leqslant 56$$
$$14x + 8y \leqslant 130$$
$$y \leqslant 6$$
$$x, \quad y \geqslant 0$$

6.4. Solve, by use of the vector method, the linear programming problem given in Problem 6.3.

6.5. Pleasure Boats, Inc., employs 180 people, 80 of whom are skilled workers; the rest are unskilled. It takes 8 skilled and 16 unskilled workers to make a sailboat. It takes 16 skilled and 12 unskilled workers to make a motorboat. The company makes a profit of $50 on each sailboat and $40 on each motorboat. Use the vector method to determine how the company can most profitably employ its workers.

7

The Simplex Method

7.1 INTRODUCTION

Among the various methods of solving linear programming problems, the simplex method is one of the most general and powerful. Actually, the graphical method, the algebraic method, and the vector method were presented primarily to give the reader a "feel" for linear programming problems and to acquaint him with some of the technical terminology so essential in understanding the rationale and mechanics of the simplex method. Otherwise in actual practice, linear programming problems of any significance are usually solved by application of the simplex method, or some variant of the simplex method.* In this chapter we illustrate the simplex method by applying it to the maximization problem of Table 7.1.

7.2 RATIONALE FOR THE SIMPLEX METHOD

The simplex method rests on two pillars: feasibility and optimality. The search for the optimal solution starts from a basic feasible solution or program.† The solution is tested for optimality, and if it is optimal, the search is

---

* Revised simplex or dual simplex are two such variants. See Loomba and Turban [1974].
† As in previous methods, we search only the extreme-point solutions.

stopped. If the test of optimality shows that the current solution is not optimal, a new and *better* basic feasible solution or program is designed. The feasibility of the new solution is guaranteed by the mechanics of the simplex method, as is the fact that each successive solution is designed only if it is better than each of the previous solutions.* This iterative process is continued until an optimal solution has been designed. As compared to the graphic, algebraic, and vector methods, the simplex method yields considerable reductions in computational requirements.

The simplex method is based on the property that the optimal solution to a linear programming problem, if it exists, can always be found in one of the basic feasible solutions.† Thus, in the simplex method, the first step is always to obtain a basic feasible solution. As explained in the vector method (Chapter 6), this means that we obtain a set of basic vectors and a set of nonbasic vectors. Further, this means that the constant or requirement vector $\mathbf{P}_0$ is expressed as a linear combination of a set of the basic vectors. This, in effect, gives us a solution to the problem. This solution is then tested for optimality by examining the net effect on the linear objective function of introducing one of the nonbasic vectors to replace at least one of the current basic vectors. If any improvement potential is noted, the replacement is made, always by introducing *only one* nonbasic vector at a time. As explained in Chapter 6, this replacement results in a new basis.

We have thus far not indicated anything that is different from the vector method. However, as we shall see below, the beauty of the simplex method lies in the fact that the relative exchange profitabilities (net effect) of all the nonbasic vectors can be determined simultaneously‡ and easily. Furthermore the replacement process is such that the new basis does not violate the feasibility of the solution.

The simplex method is quite simple and mechanical in nature. The iterative steps of the simplex method are repeated until a finite optimal solution, if it exists, is determined. Otherwise, the method indicates either that the given linear programming problem has no solution or that no finite solution exists.

7.3 THE SIMPLEX METHOD—AN ILLUSTRATIVE EXAMPLE

To fix ideas and to facilitate comparisons with the other methods of solving linear programming problems, we shall solve the problem of Table 3.2 by the simplex method. For quick reference, the data of Table 3.2 are reproduced as Table 7.1.

* Improvement between successive solutions is indicated by the signs of the entries in the net evaluation row ($C_j - Z_j$). See Section 7.3.2.

† If the problem is degenerate, the optimal solution is a degenerate basic feasible solution.

‡ In the vector method as presented in Chapter 6, the net effect of each nonbasic vector was determined individually.

Table 7.1

| Department | Product | | Capacity |
	A	B	
Cutting	10	6	2,500
Folding	5	10	2,000
Packaging	1	2	500
Profit per Unit	$23	$32	

As in the other methods, our first step is to translate the technical data into inequalities; convert these inequalities into equations by the addition of slack* variables; express the nonnegativity constraints and state the corresponding linear objective function. The slack variable may be given the familiar physical interpretation in which the capacities of cutting, folding, and packaging departments not utilized in producing products X and Y are, respectively, used to produce imaginary products S_1, S_2, and S_3, each giving a per unit profit contribution of zero.

Our problem, then, can be stated as follows:

Maximize $\quad 23X + 32Y + 0S_1 + 0S_2 + 0S_3$

subject to the constraints

$$10X + 6Y + 1S_1 + 0S_2 + 0S_3 = 2,500$$
$$5X + 10Y + 0S_1 + 1S_2 + 0S_3 = 2,000 \quad\quad (7.1)$$
$$1X + 2Y + 0S_1 + 0S_2 + 1S_3 = 500$$

and

$$X, \quad Y, \quad S_1, \quad S_2, \quad S_3 \geqslant 0$$

The simplex method, which, in effect, is a concentrated and more efficient arrangement of the algebraic and vector methods, proceeds to solve the problem by designing and redesigning successively better basic feasible solutions until an optimal solution is obtained. Each program, as we shall see below, is given in the form of a matrix or tableau. Although there are various forms for a simplex tableau, we shall follow the one given in Figure 7.1, which contains an initial program for the problem given in Table 7.1. The nomenclature of the simplex tableau, as identified in Figure 7.1, will be followed

* If the inequalities are of the "greater than or equal to" type, the surplus variables are subtracted from the left sides of the inequalities. Strict equalities require the addition of an artificial variable whose objective function coefficient represents a huge penalty. See Section 7.5.4.

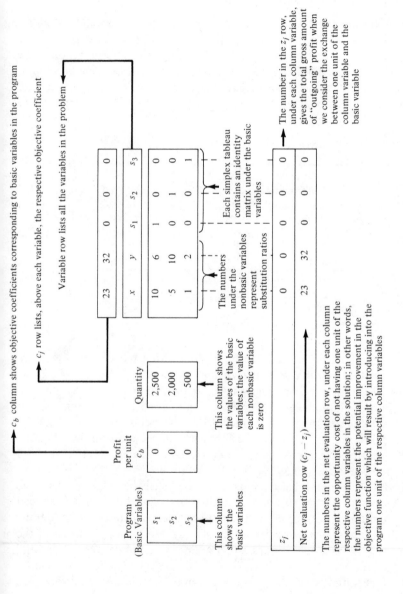

Figure 7.1

throughout this book. If the rows and columns of a given tableau are labeled in a different fashion, it is solely for the sake of convenience. To interpret the basic role and significance of these, we can always refer back to Figure 7.1.

7.3.1 Design an Initial Program

As in the algebraic and vector methods, the first program in the simplex method is that which involves only the slack variables. This program is summarized in the first simplex tableau (Table 7.2). The interpretation of the data in Table 7.2 must be fully grasped in order that the simplex method can be understood. Let us, therefore, discuss the contents of the simplex tableau shown in Table 7.2. Other simplex tableaus will have similar interpretations.

Table 7.2. First Program

Program (Basic Variables)	Profit per Unit C_b	Quantity	$C_j \rightarrow$ 23 X	32 Y	0 S_1	0 S_2	0 S_3
S_1	0	2,500	10	6	1	0	0
S_2	0	2,000	5	10	0	1	0
S_3	0	500	1	2	0	0	1
Z_j			0	0	0	0	0
$C_j - Z_j$			23	32	0	0	0

1. In the column labeled "Program" are listed the variables that are included in the solution (products being produced). These are the basic variables. Thus, in our first program, we are producing only S_1, S_2, and S_3.

2. In the column labeled "Profit per Unit" are listed the profit coefficients of the basic variables that are included in the specific program. Thus the profit coefficients of S_1, S_2, and S_3, which are included in the initial program, are listed in this column. As can be ascertained from the objective function in (7.1), the coefficients of S_1, S_2, and S_3 are all zero.

3. In the column labeled "Quantity" are listed the values of the basic variables included in the solution (quantities of the products being produced in the program). Since our initial program consists of producing 2,500 units of S_1, 2,000 units of S_2, and 500 units of S_3, these values are listed in the "Quantity" column. Any variables that are not listed under the "Program" column are the nonbasic variables. Their values are, by definition, zero. Hence in view of the initial program of Table 7.2, X and Y are the nonbasic variables, and each has a value of zero.

4. The total profit contribution resulting from a specific program can be calculated by multiplying corresponding entries in the "Profit per Unit"

column and the "Quantity" column and adding the products. Thus total profit contribution in our first program is zero.

$$2,500(0) + 2,000(0) + 500(0) = 0$$

5. Numbers in the main body (entries in columns X and Y) can be interpreted to mean physical ratios of substitution if the program consists only of the slack variables. These physical ratios of substitution* in Table 7.2 correspond exactly to the given technical specifications. For example, the number 10 gives the rate of substitution between X and S_1.† In other words, if we wish to produce 1 unit of X, 10 units of S_1 must be sacrificed. That is, available cutting capacity will be reduced by 10 units. The numbers 5 and 1 have similar interpretations. By the same token, to produce 1 unit of Y, we must "sacrifice" 6 units of S_1, 10 units of S_2, and 2 units of S_3.

6. Like the numbers in the main body, the entries in the "identity" (columns S_1, S_2, and S_3 in Table 7.2) can be interpreted as ratios of exchange. Thus the numbers in columns S_1 represent, respectively, the ratios of exchange between S_1 and the basic variables S_1, S_2, and S_3. Note that the numbers under the basic variable columns (in Table 7.2 they are S_1, S_2, and S_3) *always* consist of unit vectors. Further, the unit vector is so constructed that the number 1 is at the intersection of the row and column of the same basic variable, while zeros appear at the intersection of the different basic variables. This assertion can be checked in any simplex tableau.

7. The C_j numbers at the top of the columns of all the variables represent the coefficients of the respective variables in the objective function.

8. The numbers in the Z_j row, under each variable, give the total gross amount of outgoing profit when we consider the exchange between one unit of column variable and the basic variables.

9. The numbers in the net evaluation row, $C_j - Z_j$,‡ give the *net* effect of exchange between each variable and basic variables. They are always zero under the basic variables. Thus they are zero under S_1, S_2, and S_3 in Table 7.2. Under the nonbasic variables, they can be positive, negative, or zero.

7.3.2   Test for Optimality

In so far as the total profit contribution resulting from our initial program is zero, it can obviously be improved and hence is not the optimal program. However, the test of optimality can be formalized in terms of the signs of the entries $(C_j - Z_j)$ in the net evaluation row. In maximization problems, the program is optimal if each $C_j - Z_j$ is either zero or negative.

---

* In subsequent tableaus, these numbers are algebraic ratios of exchange.

† Note that 10 lies at the intersection of column X and row S_1.

‡ C_j is the profit coefficient of any variable j; Z_j is the sum of the products of the exchange ratios under the jth variable and the corresponding profit coefficients of the basic variables.

In minimization problems, the program is optimal if each $C_j - Z_j$ is zero or positive. Let us explain.

Assume that we wish to change the program in Table 7.2 by introducing (producing) 1 unit of Y. This would, as explained previously, involve sacrificing 6 units of S_1, 10 units of S_2, and 2 units of S_3. The net effect of this exchange on the profit function would be

$$\boxed{+1(32)} - \boxed{6(0) + 10(0) + 2(0)} = 32 = \boxed{C_j - Z_j}$$

C_j = incoming unit Z_j = outgoing total   net effect
    profit   profit

In other words, the introduction of 1 unit of Y at this stage of the solution will increase the value of the profit function by \$32. Thus the *opportunity cost** of not having this unit of Y in our initial program is \$32. It is this number, $C_j - Z_j$, that is entered in the net evaluation row under column Y. Similarly, as can easily be ascertained, the opportunity cost of not having product X in our solution, at this stage, is \$23 per unit. This is the significance of the $C_j - Z_j$ numbers in the net evaluation row. The mechanics of calculating the $C_j - Z_j$ in any tableau is as follows.

To get a number $C_j - Z_j$, under any column j, first multiply the exchange (or substitution) ratios in that column by the corresponding profit coefficients of the basic variables that appear under the C_b column; add the products (to yield Z_j) and then subtract this sum from the corresponding number listed in the C_j row.†

The numbers $C_j - Z_j$ in the net evaluation row represent the potential improvement in the objective function that will result from the introduction into the program of 1 unit of each of the respective variables. Thus, by definition, these numbers represent the opportunity costs of *not* having 1 unit of each variable in the solution. Since we are dealing with a linear programming model that assumes certainty, the presence of any positive opportunity cost (or positive $C_j - Z_j$ in maximization problems) in the net evaluation row of a given simplex tableau indicates that an optimal solution has not been obtained, and therefore a better program can be designed. This is the test that will be used in this book for obtaining an optimal solution to a maximization problem.

By use of a similar argument it can be shown that the test of optimality for minimization problems requires that each $C_j - Z_j$ be either positive or zero. We summarize next the criteria for tests of optimality.

---

* Opportunity cost is the cost involved in not following the best course of action.
† Mathematically,

$$C_j - Z_j = C_j - \sum_{i=1}^{m} a_{ij} C_i, \qquad j = 1, 2, \ldots, n \qquad (7.2)$$

where a_{ij} is the substitution ratio at the intersection of the ith row and jth column.

For maximization problems: If one or more $C_j - Z_j$ are positive, the solution is *not* optimal. If each $C_j - Z_j$ is either negative or zero, the solution *is* optimal.

For minimization problems: If one or more $C_j - Z_j$ are negative, the solution is *not* optimal. If each $C_j - Z_j$ is either positive or zero, the solution *is* optimal.

The $C_j - Z_j$ for Table 7.2 are calculated and listed in the net evaluation row. An examination of these numbers shows that they are either positive or zero. Hence our first program is not an optimal program and must be revised.

7.3.3 Revise the Current Program

Once a program is found to be nonoptimal, we must design a new and better program. How the simplex method accomplishes this task is explained below.

1. *Identify the incoming variable (or the key column).* The two positive $C_j - Z_j$ (23, 32) in the net evaluation row of Table 7.3 indicate, respectively, the magnitudes of the opportunity costs of not including 1 unit of variables (products) *X* and *Y*, at this solution stage. Since the highest opportunity cost falls under column *Y*, the variable (product) *Y* should be brought into the new program first. Hence *Y* is the incoming variable, and column *Y* is called the *key column.** The rule for determining the key column is

> *The key column is the column under which the largest positive*
> $C_j - Z_j$ *appears.*

2. *Identify the outgoing variable (or the key row†).* After we have decided to bring in variable (product) *Y* to replace at least one of the basic variables (products) in the current program (S_1, S_2, or S_3), the question becomes: How many units of *Y* can be brought in without exceeding the existing capacity of any one of the resources? In linear programming terms, this means that we must calculate the maximum allowable number of units of *Y* that can be brought into the program without violating the nonnegativity constraints. If we examine Table 7.3, we note that to bring in 1 unit of *Y*, we must sacrifice 6 units of S_1, 10 units of S_2, and 2 units of S_3. Insofar as we are currently producing only 2,500 units of S_1, it is clear that no more than $416\frac{2}{3}$ units ($\frac{2,500}{6} = 416\frac{2}{3}$) of *Y* can be brought in without violating the capacity constraint of the cutting department. Similarly, at this stage of the solution, the production of *Y* is limited to 200 units ($\frac{2,000}{10} = 200$) and 250 units ($\frac{500}{2} = 250$) by the available capacities of the folding department and packaging department, respectively. The limiting case, therefore, arises from row S_2 in Table 7.3. This, then, is our key row (the outgoing variable) and 200 units is the maximum quantity of the product *Y* that can be produced,

---

* Also referred to in the literature as the *pivot column.* See Appendix E.
† Also referred to in the literature as the *pivot row.* See Appendix E.

Table 7.3. First Program

Program (Basic Variables)	Profit per Unit C_b	Quantity	$C_i \to$ 23 X	32 Y	0 S_1	0 S_2	0 S_3	Replacement Quantity
S_1	0	2,500	10	6	1	0	0	$416\frac{2}{3}$
S_2	0	2,000	5	10	0	1	0	200 → outgoing variable
S_3	0	500	1	2	0	0	1	250
Z_j			0	0	0	0	0	
$C_j - Z_j$			23	32	0	0	0	

incoming variable

at this stage of the solution, without violating the nonnegativity constraints. The mechanics for identifying the key row will now be discussed.

Divide the entries under the "Quantity" column by the corresponding *positive** entries of the key column, and compare these ratios. *The row in which the smallest ratio falls is the key row.* These ratios are, in effect, maximum possible *replacement quantities* that indicate the limiting values of the incoming variable as a replacement for each of the basic variables. Hence from now on we shall refer to them as, and list them under, a column named "Replacement Quantity."

The calculations that help identify the outgoing variable or the key row are as follows.

For the cutting department (row S_1):

$$\frac{2,500}{6} = 416\tfrac{2}{3} \text{ units}$$

For the folding department (row S_2):

$$\frac{2,000}{10} = 200 \text{ units} \rightarrow \text{key row}$$

For the packaging department (row S_3):

$$\frac{500}{2} = 250 \text{ units}$$

While going through the simplex algorithm, it is convenient to place these replacement-quantity calculations on the extreme right-hand side of a given tableau. The limiting replacement quantity of the incoming variable (product) is then identified by the *lowest nonnegative* value.† The outgoing variable or the key row is indicated by an arrow of the form →, as shown in Table 7.3.

3. *Identify the Key Number* (*or the Pivot‡ Element*). Once the key row and the key column have been determined, the identification of the key number is a simple matter. *The number that lies at the intersection of the key row and the key column of a given tableau is the key number.* Thus the key number in Table 7.3 is 10. The identification of the key column and the key row has shown us that the variable (product) Y will replace variable (product) S_2 and that no more than 200 units of Y can be produced under the current capacity restrictions. Our next task is to determine the exact composition of the remainder of the revised program. In other words, we must find the reductions in S_1 and S_3 due to the fact that 200 units of Y are to be included

---

* A negative entry in the key column, when interpreted as a ratio of exchange, would mean that the introduction of the incoming variable increases rather than decreases the magnitude of the outgoing variable in which this negative entry exists. The current magnitude of this outgoing variable, therefore, would provide no limit to the introduction of the incoming variable. Similarly, an exchange ratio of zero would provide no limit. Hence, in identifying the key row, only the positive ratios of substitution need be examined.

† Note that a replacement quantity of "zero" would always constitute the lowest nonnegative value.

‡ See Appendix E.

in the revised program. Furthermore, we must build an entire new simplex tableau for the revised program.

Since it takes 10 units of folding-department capacity to produce 1 unit of Y, it is evident that all 2,000 units (200 × 10) of the folding capacity are exhausted. However, as we noted earlier, the production of 1 unit of Y also requires 6 units of cutting-department capacity and 2 units of packaging-department capacity. Thus the remaining capacity of the cutting department is $2,500 - (200 \times 6) = 1,300$, and the remaining capacity of the packaging department is $500 - (200 \times 2) = 100$ units.

Another way to say the same thing is that our second program calls for producing $S_1 = 1,300$, $Y = 200$, $S_3 = 100$; and $X = 0$, $S_2 = 0$. The three products (S_1, Y, and S_3) included in the second program are listed in the "Program" column in Table 7.4. This second program here corresponds to point M in Figure 3.6 and to using $\mathbf{P}_3$, $\mathbf{P}_2$, and $\mathbf{P}_5$ as the basic vectors in the vector method (Section 6.4.3). It also corresponds to Equations (4.12) to (4.14) in the algebraic method (Section 4.4.4). Further, it should be emphasized that the ratios of substitution among different variables at this stage of the solution are those that were obtained in Equations (4.12) to (4.14) of the algebraic method. Since a specific simplex tableau contains these very ratios of substitution, we could use such information as given in Equations (4.12) to (4.14) to build Table 7.4.* The simplex method, however, provides mechanics with which the task can be accomplished simply. Let us illustrate.

Table 7.4, which will contain our second program, is to be derived from Table 7.3, which represented our initial program. The simplex tableaus are so constructed that the number of rows in each tableau is the same, even though in some cases the values of one or more basic variables appear as zero under the "Quantity" column. Further, any given tableau, during the solution stages, has two types of rows: (1) the key row, and (2) the nonkey rows. Thus to derive a new tableau from an old tableau, all we have to do is to establish rules of transformation for these two types of rows. These rules of transformation form the mechanical foundations of the simplex method. The rules are to be applied to the entire set of entries of each row, starting with and to the right of the "Quantity" column.

Transformation of the Key Row

The rule for transforming the key row is

Divide all the numbers in the key row by the key number. The resulting numbers form the corresponding row in the next tableau (to be placed in exactly the same position).

---

* Equations (4.12) to (4.14) are rearranged in the following form:

$$1,300 = 7X + 0Y + 1S_1 - 0.6S_2 + 0S_3 \tag{4.13}$$
$$200 = 0.5X + 1Y + 0S_1 + 0.1S_2 + 0S_3 \tag{4.12}$$
$$100 = 0X + 0Y + 0S_1 - 0.2S_2 + 1S_3 \tag{4.14}$$

Table 7.4. Second Program

Program (Basic Variables)	Profit per Unit C_b	Quantity	$C_j \to$ 23 X	32 Y	0 S_1	0 S_2	0 S_3	Replacement Quantity
S_1	0	1,300	7	0	1	$-\frac{3}{5}$	0	$\frac{1,300}{7} \to$ outgoing variable
Y	32	200	$\frac{1}{2}$	1	0	$\frac{1}{10}$	0	400
S_3	0	100	0	0	0	$-\frac{1}{5}$	1	
Z_j			16	32	0	$\frac{16}{5}$	0	
$C_j - Z_j$			7	0	0	$-\frac{16}{5}$	0	

incoming variable

161

Thus the second row in Table 7.4 (row Y) is derived from the second row in Table 7.3 (row S) by simply dividing all the numbers in row S_2 by 10 (the key number). The new row Y (Table 7.4) is

$$200 \quad \tfrac{1}{2} \quad 1 \quad 0 \quad \tfrac{1}{10} \quad 0$$

Transformation of the Nonkey Rows

The rule for transforming a nonkey row is

Subtract from the old row number (in each column) the product of the corresponding key-row number and the corresponding fixed ratio (formed by dividing the old row number in the key column by the key number). The result will give the corresponding new row number (to be placed in exactly the same position).

The above rule can be placed in the following equation form:

new row number = old row number

$$- \left(\begin{array}{c} \text{corresponding number} \\ \text{in key row} \end{array} \times \begin{array}{c} \text{corresponding} \\ \text{fixed ratio} \end{array} \right)$$

where

$$\text{fixed ratio} = \frac{\text{old row number in key column}}{\text{key number}}$$

Thus the new row S_1 for Table 7.4 is derived as follows (corresponding fixed ratio $= \tfrac{6}{10} = 0.6$):

old row number	$-$	corresponding number in old key row	$\times$	corresponding fixed ratio	$=$	new row number
2,500	$-$	2,000	$\times$	0.6	$=$	1,300
10	$-$	5	$\times$	0.6	$=$	7
6	$-$	10	$\times$	0.6	$=$	0
1	$-$	0	$\times$	0.6	$=$	1
0	$-$	1	$\times$	0.6	$=$	-0.6
0	$-$	0	$\times$	0.6	$=$	0

Similarly, the new S_3 row in Table 7.4 is derived as follows (corresponding fixed ratio $= \tfrac{2}{10} = 0.2$):

old row number	$-$	corresponding number in old key row	$\times$	corresponding fixed ratio	$=$	new row number
500	$-$	2,000	$\times$	0.2	$=$	100
1	$-$	5	$\times$	0.2	$=$	0
2	$-$	10	$\times$	0.2	$=$	0
0	$-$	0	$\times$	0.2	$=$	0
0	$-$	1	$\times$	0.2	$=$	-0.2
1	$-$	0	$\times$	0.2	$=$	1

The results are entered in the second simplex tableau, shown in Table 7.4. As indicated in the table, our second program calls for the production of $S_1 = 1,300$, $Y = 200$, and $S_3 = 100$ units. The variables X and S_2 are not in the solution (program) and, therefore, assume values of zero. The total profit contribution resulting from this program is $0(1,300) + 32(200) + 0(100) = \$6,400$.

The $C_j - Z_j$ numbers of the net evaluation row are calculated as before, according to Equation (7.2), and are shown in Table 7.4.*

7.3.4 Design Another Improved Program

As the net evaluation row of Table 7.4 shows, we still have one positive $C_j - Z_j$ under the X column. Its value of $+7$ indicates a positive opportunity cost of not having 1 unit of X in the program. Hence program 2, shown in Table 7.4, is not an optimal program. This calls for designing a new program and, therefore, deriving a new simplex tableau. The procedure for deriving this third simplex tableau (Table 7.5) is exactly the same as was followed in deriving the second simplex tableau. Before deriving the third tableau, let us emphasize again that the numbers contained in the rows of this tableau will be the same as the coefficients of the different variables in the equations that represents program 3 in the algebraic method.† This type of correspondence can be identified not only between the simplex method and the algebraic method but essentially among all the methods of solving linear programming problems discussed in this book. The advantage of the simplex method lies in the easy mechanical nature of its solution process, whereby a finite number of iterations takes us to the optimal solution. Important as the simplex method is for learning purposes, we can really appreciate its value by noting that this type of procedure can easily be programmed into a computer.‡ Thus linear programming problems of even very large dimensions can be solved in relatively short periods of time.

Derivation of Table 7.5

Were there more than one positive $C_j - Z_j$ numbers in the net evaluation row of a simplex tableau, we would select the largest one to identify the key column (the incoming variable). Since in Table 7.4 there is only one positive $C_j - Z_j$, we choose X (the column under which the positive $C_j - Z_j$ appears) to be the incoming product. Thus the column labeled X is the key column. As before, we must now determine the limit on the quantity of X

---

* The $C_j - Z_j$ row can also be treated as a nonkey row and can, therefore, be transformed from one tableau to the next by following the rules of transformation for the nonkey row. The reader can verify this statement by calculating the net evaluation row of Table 7.4 from that of Table 7.3.

† Rearrange the equations so that the constant terms are on the left-hand side.

‡ See Section 12.2.3 for an illustration.

that can be introduced into the program and thus identify the key row. The maximum replacement quantities of variables S_1 and Y are shown on the right-side of Table 7.4 under the "Replacement Quantity" column.* They indicate that row S_1 is the key row, and hence 7 is the key number.

By the rules of transformation presented earlier, the first row of Table 7.5 (row X) is derived from the first row of Table 7.4 (row S_1) by simply

Table 7.5. Optimal Program

Program (Basic Variables)	Profit per Unit C_b	Quantity	$C_j \rightarrow$ 23 X	32 Y	0 S_1	0 S_2	0 S_3
X	23	$\frac{1,300}{7}$	1	0	$\frac{1}{7}$	$-\frac{3}{35}$	0
Y	32	$\frac{750}{7}$	0	1	$-\frac{1}{14}$	$\frac{1}{7}$	0
S_3	0	100	0	0	0	$-\frac{1}{5}$	1
Z_j			23	32	1	$\frac{13}{5}$	0
$C_j - Z_j$			0	0	-1	$-\frac{13}{5}$	0

dividing all the numbers in row S_2 by 7 (the key number). Row X of Table 7.5 becomes

$$\frac{1,300}{7} \quad 1 \quad 0 \quad \tfrac{1}{7} \quad -\tfrac{3}{35} \quad 0$$

The nonkey rows of Table 7.4 are transformed as follows:

$$\frac{\text{old row number}}{\text{number}} - \left(\begin{array}{c} \text{corresponding number} \\ \text{in old key row} \end{array} \times \begin{array}{c} \text{corresponding} \\ \text{fixed ratio} \end{array} \right) = \frac{\text{new row}}{\text{number}}$$

The *calculation for row Y* (fixed ratio $= \frac{1}{14}$) is

$$200 - 1,300 \times \tfrac{1}{14} = \tfrac{750}{7}$$
$$\tfrac{1}{2} - 7 \times \tfrac{1}{14} = 0$$
$$1 - 0 \times \tfrac{1}{14} = 1$$
$$0 - 1 \times \tfrac{1}{14} = -\tfrac{1}{14}$$
$$\tfrac{1}{10} - -\tfrac{3}{5} \times \tfrac{1}{14} = \tfrac{1}{7}$$
$$0 - 0 \times \tfrac{1}{14} = 0$$

Since the fixed ratio for row S_3 (Table 7.4) is zero, the corresponding row in the next program (Table 7.5) is exactly the same.

All the $C_j - Z_j$ numbers in the net evaluation row of Table 7.5 are now either negative or zero, indicating the optimality of Program 3.

---

* We need not check row S_3 because its ratio of substitution (under the key column) is zero. Remember that replacement quantities need to be calculated only for positive ratios of exchange (substitution).

7.4 ECONOMIC INTERPRETATION OF THE $C_j - Z_j$ NUMBERS IN THE OPTIMAL PROGRAM

The economic interpretation of the $C_j - Z_j$ numbers in the net evaluation row of the optimal program is helpful for managerial decisions. Since the net evaluation of S_1, at this stage, is -1.0, the introduction of 1 unit of S_1 (letting 1 unit of the cutting-department capacity stay idle) will decrease the objective function by \$1. By the same reasoning, if we had 1 more unit of the cutting capacity, the objective function could be increased by \$1. In other words, \$1 gives us the *marginal worth, artificial accounting price, or shadow price* of 1 unit of cutting capacity. Information regarding the marginal worths of various resources can help the manager decide whether additional resources should be purchased and at what prices.

The term *marginal worth* conveys the idea of the economists' concept of the worth of a marginal unit of a given resource. Here, as can be seen from the $C_j - Z_j$ row (under column S_2) of Table 7.5, the marginal worth of the folding capacity is \$2.60.* The shadow price of the packaging capacity, however, is zero (see the $C_j - Z_j$ value under S_3 in Table 7.5). This makes economic sense because the optimal tableau indicates that we still have a spare capacity of 100 units in the packaging department ($S_3 = 100$). Hence its marginal worth must be zero. The value of all the available resources can be calculated by multiplying the given capacity levels of the different resources by their respective marginal worths and adding the products. In our case, this value is equal to $\$7,700 = 2,500(1) + 2,000 (2.60) + 500(0)$. Comparing this *imputed* value of the available resources with the value of the objective function in the optimal program (Table 7.5), we find that their magnitudes are exactly the same. The fact that the value of the objective function in the optimal program equals the *imputed* value of the available resources is one component of a linear programming problem called the dual problem (Chapter 8).

7.5 THE SIMPLEX METHOD (MINIMIZATION CASE)

7.5.1 Preliminaries

One way to solve a minimization problem is to convert it to a maximization problem and then solve it by the simplex algorithm illustrated in Section 7.4. The conversion from minimization to maximization (or vice versa) can be made by just changing the signs of the cost (or profit) coefficients in the objective function. The constraints are left untouched. The solution to

---

* The same information can be obtained by examining the final form of the objective function in the algebraic method. See Equation (4.18) in Section 4.4.5.

the converted problem is the same as that to the original problem, except that the sign of the optimal value of the objective function of the converted problem is changed to obtain the optimal value of the objective function of the original problem.*

The assertion of the above paragraph can be verified by taking the minimization problem of Table 7.6, converting it to a maximization problem (change the signs of the objective function coefficients), solving it by the simplex method, and then comparing the results with the solution to the original minimization problem, shown in Table 7.9. Figure 7.2 shows the optimal corner point C, which corresponds to $X = 15$ and $Y = 2.5$.

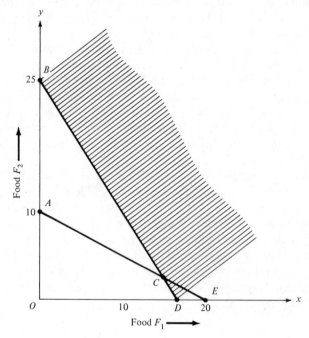

Figure 7.2

Our purpose in this section is to illustrate the straightforward application of the simplex method to minimization problems. The simplex algorithm that we illustrated in the maximization case in Section 7.4 is also applicable to minimization problems. There are, however, two specific differences. The first relates to the test of optimality and hence to the choice of the incoming variable. As described earlier, in the maximization case the optimality is established when all $C_j - Z_j$ are either negative or zero (and hence the variable with the largest positive $C_j - Z_j$ is chosen as the incoming

* $\min CX = -\max(-C)X$. Also, $\max CX = -\min(-C)X$.

variable). Conversely, in the minimization case, the optimality is established when all $C_j - Z_j$ are either positive or zero (and hence the variable with the largest negative $C_j - Z_j$ is chosen as the incoming variable). The second difference is the use of artificial variables as the means of obtaining the initial basic feasible solution. Let us illustrate.

7.5.2   The Problem

The technical specifications of a diet-type problem are given in Table 7.6. The data indicate that 1 unit of F_1 (say, 1 ounce) contains 2 units of vitamin A and 3 units of vitamin B. Similarly, 1 unit of F_2 contains 4 units of vitamin A

Table 7.6

	Food		Daily Requirement
Vitamin	F_1	F_2	
A	2	4	40
B	3	2	50
Cost per Unit (cents)	3	2.5	

and 2 units of vitamin B. The daily requirement for vitamin A is at least 40 units, and for vitamin B at least 50 units.* Our objective is to determine optimal quantities of foods F_1 and F_2 to be purchased so that the daily vitamin requirements are met and, simultaneously, the cost of purchasing the foods is minimized. Assuming that X represents the quantity of food F_1 and Y the quantity of food F_2 to be purchased, we may state the problem in algebraic terms as follows:

Minimize $3X + 2.5Y$

subject to the constraints

$$2X + 4Y \geqslant 40$$
$$3X + 2Y \geqslant 50$$

and

$$X, \quad Y \geqslant 0$$

* Any intake of vitamins in excess of the daily requirements is assumed not to be harmful.

7.5.3 Transforming the Inequalities into Equalities

As opposed to those in the problem discussed in Section 7.3, the structural constraints here are of the "greater than or equal to" type. Hence converting the inequalities into equalities requires the subtraction of surplus rather than the addition of slack variables. Let variables R_1 and R_2 represent, respectively, the quantity of vitamin A in excess of 40 units and the quantity of vitamin B in excess of 50 units. The introduction of these surplus variables converts the above inequalities into the following equations:

$$2X + 4Y - R_1 \qquad = 40 \qquad\qquad (7.3)$$

$$3X + 2Y \qquad - R_2 = 50 \qquad\qquad (7.4)$$

Variables R_1 and R_2 have the property that 1 unit of R_1 contains 1 unit of vitamin A and 1 unit of R_2 contains 1 unit of vitamin B. Variable R_1 appears only when we buy foods F_1 and F_2 with the purpose of satisfying the requirement for vitamin A, and R_2 appears only when we buy foods F_1 and F_2 to satisfy the requirement for vitamin B. Thus, if a specific program of purchasing foods F_1 and F_2 is such that the presence of R_1 and R_2 is required to satisfy Equations (7.3) and (7.4), the magnitudes of R_1 and R_2 will represent, respectively, the quantity of vitamin A in excess of 40 units and the quantity of vitamin B in excess of 50 units.

With the above interpretation, it is easy to see that variables R_1 and R_2 are restricted to nonnegative values, and each has a cost coefficient of zero. A complete statement of the problem, therefore, is:

Minimize $\qquad\qquad 3X + 2.5Y + 0R_1 + 0R_2$

subject to the constraints

$$2X + \quad 4Y - 1R_1 + 0R_2 = 40 \qquad\qquad (7.5)$$

$$3X + \quad 2Y + 0R_1 - 1R_2 = 50 \qquad\qquad (7.6)$$

and

$$X, \qquad Y, \quad R_1, \quad R_2 \geqslant 0$$

7.5.4 Artificial Slack Variables

Let us examine Equation (7.5). If we let real variables X and Y equal zero, we obtain a value of -40 for the surplus variable R_1. Any such negative value is unacceptable because it violates the nonnegativity constraint for R_1 and does not make any sense in terms of physical interpretation. We face a similar difficulty in connection with Equation (7.6). Hence the initial program cannot be designed by letting X and Y equal to zero. Some further modification of the structural constraints must be made. We propose, therefore, at this stage, to modify the statement of our problem in such a way that we can make X and Y equal to zero in the above equations and still have positive-valued variables that satisfy these equations. This is accomplished by introducing into the original inequalities, in addition to the surplus variables, the *artificial variables*. The artificial variables will be represented by the capital letter A with the proper subscript. Thus we can modify Equations (7.5) and (7.6) with the addition of artificial variables A_1 and A_2, respectively:

$$2X + 4Y - 1R_1 + 0R_2 + A_1 = 40 \qquad (7.7)$$

$$3X + 2Y + 0R_1 - 1R_2 + A_2 = 50 \qquad (7.8)$$

In this problem the artificial variables A_1 and A_2 can be thought of as imaginary foods, each unit containing 1 unit of the pertinent vitamin. For example, we can assume here that 1 unit of A_1 contains 1 unit of vitamin A, whereas one unit of A_2 contains 1 unit of vitamin B. In this sense A_1 is similar to R_1, and A_2 is similar to R_2. Also, both A_1 and A_2 are restricted to nonnegative values for obvious reasons. However, the correspondence between the surplus and artificial variables does not hold in the matter of cost coefficients. Whereas surplus variables have zeros as their cost coefficients, each artificial slack variable is assigned an infinitely large cost coefficient (usually denoted by M).

Thus the addition of the surplus and artificial variables converts the original problem to the following:

Minimize $3X + 2.5Y + 0R_1 + 0R_2 + MA_1 + MA_2$

subject to the constraints

$$2X + 4Y - 1R_1 + 0R_2 + 1A_1 + 0A_2 = 40 \qquad (7.9)$$

$$3X + 2Y + 0R_1 - 1R_2 + 0A_1 + 1A_2 = 50 \qquad (7.10)$$

and

$$X, \quad Y, \quad R_1, \quad R_2, \quad A_1, \quad A_2 \geqslant 0$$

If in Equations (7.9) and (7.10) we let variables X, Y, R_1, and R_2 assume values of zero, the artificial variables A_1 and A_2 will have positive values. We can see, therefore, that the inclusion of artificial variables will permit us to design an initial program in which no units of foods F_1 and F_2 are purchased, and yet the nonnegativity constraints are not violated.*

In other words, these artificial variables enable us to make a convenient and correct start in obtaining an optimal solution by the simplex method. Further, having attached to each artificial variable an extremely large cost coefficient M, we can be certain that these variables can never enter into the optimal solution.† The inclusion of even 1 unit of an artificial slack

---

* The reader will recall that such an initial program, when applied to Equations (7.5) and (7.6), violated the nonnegativity constraints for variables R_1 and R_2.

† If, in any linear programming problem involving artificial slack variables, the application of the simplex method fails to remove all artificial slack variables from the solution basis, the original problem has no solution.

variable in any program would result in a prohibitive cost. The deck is stacked, so to speak, and we would *intuitively* expect a quick exit of the artificial variables from any basic feasible solution. The solution to our modified problem (modified by the inclusion of A_1, A_2, etc.) will, therefore, give us the solution to the original problem.

It should be stated here that if the original specifications of a problem are such that a given constraint is a strict equality (as opposed to an in-equality), we also modify the problem by adding an artificial slack variable whose objective function coefficient imposes a huge, and hence unacceptable, penalty. As already discussed, such a modification permits us to obtain a basic feasible program with which to start the simplex search. The artificial variable with the unacceptable high penalty will never enter the optimal program, and hence the optimal solution to the modified problem is the optimal solution to the original problem.

7.5.5   Design an Initial Program

The first program is obtained by letting each of the variables X, Y, R_1, and R_2 assume a value of zero. This means, as can be seen in Equations (7.9) and (7.10), that the initial program calls for purchasing 40 units of A_1 and 50 units of A_2. This program is given in Table 7.7.

7.5.6   Test for Optimality

The $C_j - Z_j$ numbers in the net evaluation row are calculated, as before, according to Equation (7.2). Since $C_j - Z_j$ are negative under both X and Y, our initial program is not optimal. The negatives of the entries in the net evaluation row represent the opportunity costs of not having 1 unit of each of the variables in the program. For example, the opportunity cost of not having 1 unit of Y in the solution is $-(2.5 - 6M) = 6M - 2.5$ cents, a positive opportunity cost that must be eliminated. Since a positive op-portunity cost corresponds to a negative entry in the net evaluation row in the case of a minimization problem, we can state the following decision rule for testing the optimality of a given program in a minimization problem: *As long as there exists even a single negative $(C_j - Z_j)$ number in the net evaluation row of a minimization problem, the optimal solution has not been obtained.*

7.5.7   Revise the Initial Program

As in the simplex method for solving a maximization problem, revision of the current program in the case of a minimization problem requires (1) identification of the key column, (2) identification of the key row and the key number, and (3) transformation of the key row and the nonkey rows into the new tableau that contains the revised program. The mechanics of these

Table 7.7. First Tableau

Program (Basic Variables)	Cost per Unit C_b	Quantity	$C_j \to 3$ X	2.5 Y	0 R_1	0 R_2	M A_1	M A_2	Replacement Quantity
A_1	M	40	2	4	-1	0	1	0	$10 \to$ outgoing variable
A_2	M	50	3	2	0	-1	0	1	25
Z_j			$5M$	$6M$	$-M$	$-M$	M	M	
$C_j - Z_j$			$3 - 5M$	$\frac{5}{2} - 6M$	M	M	0	0	

incoming variable

171

steps is exactly the same as in the maximization problem. However, as mentioned earlier in the minimization case, it is the largest negative entry in the net evaluation row (as opposed to the largest positive entry in the maximization case) which identifies the key column. The reason for this is obvious. In the minimization case, if the net evaluation entry under a particular column variable is negative, it is indicative of the fact that the inclusion of this variable in the new basis (by replacement of one of the current basic variables) will decrease the value of the objective function.

Since $C_j - Z_j$ for column Y is the largest negative number, we select Y as the incoming variable (key column).*

Next we must determine how many units of Y can be brought in without making either A_1 or A_2 negative. From row A_1, the maximum amount of Y that can be brought into the solution is $\frac{40}{4} = 10$ units. From row A_2, the maximum amount of Y that can be brought into the solution is $\frac{50}{2} = 25$ units. We see, therefore, that row A_1 provides the limiting case. This is, in other words, our outgoing variable (key row), and the key number is 4.

The rest of the procedure for revising the program is exactly the same as that followed in the simplex method for a maximization problem. The key row (row A_1 in this case) is transformed by dividing all its entries by the key number; the nonkey row (row A_2 in this case) is transformed according to the transformation rule for the nonkey row.† Accordingly, a new program is designed. Table 7.8 lists the second program along with other pertinent information. Since there are negative $C_j - Z_j$ in Table 7.8, the second program is not the optimal program.

7.5.8 Revise the Second Program

The net evaluation row of Table 7.8 has negative entries under columns X and R_1. Since the $C_j - Z_j$ under column X has a larger negative value than that under column R_1, the variable X should be brought into the solution next. That is, variable X is the incoming variable. Further, we see that A_2 is the outgoing variable. The new program will, therefore, consist of Y and X. Our revised program, given in Table 7.9, is derived in precisely the same manner as before, by following the rules of transformation.

---

* Insofar as M represents a very large cost, the comparative magnitudes of the net-evaluation-row entries can be ascertained simply by comparing the number of M's. Thus, in Table 7.7, the net evaluation entry $\frac{5}{2} - 6M$ is, in absolute terms, larger than $3 - 5M$. This means that, for reducing the cost function, 1 unit of Y is preferable, at this stage, to 1 unit of X.

† As the reader will recall, the transformation rule for the nonkey rows is

$$\text{old row number} - \left(\begin{array}{c}\text{corresponding number} \\ \text{in key row}\end{array} \times \begin{array}{c}\text{corresponding} \\ \text{fixed ratio}\end{array}\right) = \text{new row number}$$

where

$$\text{fixed ratio} = \frac{\text{old row number in key column}}{\text{key number}}$$

Table 7.8. Second Tableau

Program (Basic Variables)	Cost per Unit C_b	Quantity	$C_j \to$ 3 X	2.5 Y	0 R_1	0 R_2	M A_1	M A_2	Replacement Quantity
Y	2.5	10	$\frac{1}{2}$	1	$-\frac{1}{4}$	0	$\frac{1}{4}$	0	20
A_2	M	30	2	0	$\frac{1}{2}$	-1	$-\frac{1}{2}$	1	15 → outgoing variable
Z_j			$\frac{5}{4} + 2M$	2.5	$\frac{1}{2}M - \frac{5}{8}$	$-M$	$\frac{5}{8} - \frac{1}{2}M$	M	
$C_j - Z_j$			$\frac{7}{4} - 2M$	0	$\frac{5}{8} - \frac{1}{2}M$	M	$\frac{3}{2}M - \frac{5}{8}$	0	

incoming variable

Table 7.9. Third and Optimal Tableau

Program (Basic Variables)	Cost per Unit C_b	Quantity	$C_j \to$ 3 X	2.5 Y	0 R_1	0 R_2	M A_1	M A_2
Y	2.5	$\frac{5}{2}$	1	1	$-\frac{3}{8}$	$\frac{1}{4}$	$\frac{3}{8}$	$-\frac{1}{4}$
X	3	15	0	0	$\frac{1}{4}$	$-\frac{1}{2}$	$-\frac{1}{4}$	$\frac{1}{2}$
Z_j			3	$\frac{5}{2}$	$-\frac{3}{16}$	$-\frac{7}{8}$	$\frac{3}{16}$	$\frac{7}{8}$
$C_j - Z_j$			0	0	$\frac{3}{16}$	$\frac{7}{8}$	$M - \frac{3}{16}$	$M - \frac{7}{8}$

Since all the $C_j - Z_j$ in the net evaluation row of Table 7.10 are zero or positive, the optimal solution to our problem has been obtained. This optimal program assigns a value of 15 to variable X and $\frac{5}{2}$ to variable Y. In other words, this optimal program calls for purchasing 15 units of food F_1 and $\frac{5}{2}$ units of food F_2 daily, with an attendant cost of 51.25 cents. As a quick check will show, this program meets the daily requirements of vitamins A and B.

An economic interpretation of the $C_j - Z_j$ numbers, under the slack variables in the optimal tableau of a maximization problem was presented in Section 7.4. It will be recalled that, in the maximization case, the $C_j - Z_j$ number (in the optimal tableau) associated with a slack variable represented the marginal worth or imputed price per unit of a given resource. Further, we showed in Section 7.4 that the sum total of the imputed values of all the resources was exactly equal to the value of the objective function in the optimal program. A similar situation obtains for the $C_j - Z_j$ numbers in the optimal tableau of a minimization problem. In the minimization case, the $C_j - Z_j$ number (in the optimal tableau) associated with a surplus variable represents the marginal worth or imputed value that is assigned to one unit of the required item or characteristic. For example, the $C_j - Z_j$ number under R_1 in the optimal program shown in Table 7.9 is $\frac{3}{16}$. This means that we have assigned an imputed value of $\frac{3}{16}$ cent to 1 unit of vitamin A. Similarly, as can be seen from Table 7.9 (examine the $C_j - Z_j$ number under R_2), the imputed value of 1 unit of vitamin B is $\frac{7}{8}$ cent. Further, note that the sum total of the imputed values of the two stated requirements of vitamins $[\frac{3}{16}(40) + \frac{7}{8}(50) = 51.25$ cents] exactly equals the total cost of the optimal program shown in Table 7.9.

7.6 SPECIAL CASES

The usual case in linear programming problems is that we have a unique optimal solution. However, the following special cases should be noted:

(1) multiple optimal solutions,* (2) no feasible solution,† (3) unbounded solutions,‡ and (4) degenerate basic feasible solutions (degeneracy). All four special cases were stated and briefly explained in Chapter 3. We illustrate the case of degeneracy in the next section.

7.7 DEGENERACY

It will be recalled that the simplex method is based on a set of rules whereby we proceed from one basic feasible solution to the next until an optimal solution, if it exists, is obtained. Each new simplex program, if we think in terms of vectors, is obtained by choosing a new set of basic vectors. The iterative process, therefore, consists in going from the old basis to a new basis. The new basis is chosen by replacing at least one of the vectors currently in the basis with only one of the nonbasic vectors. The vector to be introduced and the vector to be replaced correspond, respectively, to the key column and the key row of the simplex method. To proceed from one solution to the next by the simplex method, as the reader will recall, requires the identification of the key column and the key row.

Selection of the key column is a simple task, for it simply involves the identification of the column that contains the largest positive entry (maximization case) or the largest negative entry (minimization case) in the net evaluation row of a simplex tableau. However, in selecting the key row for purposes of replacing one of the basic vectors, we can face two difficulties.

1. The initial simplex tableau may be such that one or more of the variables currently in the basis has a value of zero (one or more entries in the "Quantity" column is zero). If this happens, the minimum replacement quantity will be zero. It will, then, appear that the replacement process cannot be continued, for the variable to be replaced is already zero.

2. The minimum replacement quantities for two or more variables currently in the basis may be the same. If this happens, there is a tie in terms of selection of the key row. In this case, removal of one of the tied variables will also reduce the other tied variable(s) to zero. In the next simplex tableau of this case, therefore, one or more of the basic vectors will have a value of zero.

---

* If any of the $C_j - Z_j$ numbers of the nonbasic variables, in the optimal simplex tableau, are zero, we have multiple optimal solutions.

† The case of "no feasible solution" represents a problem where the set of constraints is inconsistent (i.e., mutually exclusive). In the simplex algorithm, this case will occur if the solution is optimal but some artificial variable is in the basis with a nonzero value.

‡ Think of a simplex tableau with only one unfavorable $C_j - Z_j$. This identifies the only incoming candidate. Assume now that all ratios of substitution under this key column are either zero or negative. This means that no change in basis can be made, and no current basic variable can be reduced to zero. Actually, as the incoming variable is introduced, we continue to increase, without bounds, those basic variables whose ratios of substitution are negative. This illustrates an unbounded solution.

Both the above conditions give rise to a phenomenon known as *degeneracy*. Attempts to solve a degenerate linear programming problem will show that either (1) after a finite number of iterations the optimum solution can be obtained, or (2) the problem begins to cycle,* thereby preventing the attainment of the optimal solution. What happens is that at least one of the components of the solution vector assumes a value of zero, and we have a *degenerate basic feasible* solution. Let us illustrate with the same example that was used to illustrate a degenerate basic feasible solution in Section 3.5. (See Figure 7.3.)

Maximize $\qquad\qquad 40X + 50Y$

subject to the constraints

$$25X + 20Y \leqslant 5,000$$
$$20X + 30Y \leqslant 7,500$$

and

$$X, \qquad Y \geqslant \quad 0$$

7.7.1 Design an Initial Program

Shown in Table 7.10 is the first program in which both the basic variables are the slack variables S_1 and S_2. Calculation of the net evaluation row in Table 7.10 shows that column Y is the key column. As previously, our next task is to choose a key row by identifying that variable (product) which is to be replaced by the incoming variable or product (in this case the incoming variable is Y). But there is no unique key row in Table 7.10, since both row S_1 and row S_2 provide the limiting case. In other words, we have a tie between row S_1 and row S_2.

The introduction of 250 units of Y at this stage will require removing all units of S_1 and S_2 from the solution. This means that our next program will consist of only 250 units of Y. It appears, therefore, that our next tableau would have only one row instead of the two rows contained in Table 7.10. However, in the simplex algorithm all tableaus, during all the solution stages, have the same number of rows. How, then, do we proceed in the case in which a tie appears? The answer is that, insofar as the simplex tableau format requires the replacement of only one basic variable at a time, we should somehow break the tie between row S_1 and row S_2 by designating one of them as the key row. Then, only that variable which falls in the key row should be shown as removed from the basis. The other will remain a basic variable, but its value will be zero. The mechanics for accomplishing this will be discussed in the following paragraphs.

---

* That is, during the solution stages, we keep returning to the same basis.

Table 7.10. First Tableau

Program (Basic Variables)	Profit per Unit C_b	Quantity	$C_j \rightarrow$ 40 X	50 Y	0 S_1	0 S_2	Replacement Quantity
S_1	0	5,000	25	20	1	0	250 → outgoing variable
S_2	0	7,500	20	30	0	1	250
Z_j			0	0	0	0	
$C_j - Z_j$			40	50	0	0	

incoming variable

7.7.2 Solving a Degenerate Problem

In Table 7.10 we have encountered a degenerate situation. How do we resolve the degeneracy? Some rule is obviously needed to break the tie between the two basic variables S_1 and S_2. Several arbitrary rules have been suggested for making this decision. One such rule is that the variable whose subscript is smallest* should be removed first. Another rule calls for removing that variable whose subscript is found first in a table of random numbers. Another alternative, of course, is to remove one of the tied variables at will. All these alternatives are arbitrary, but they do permit the continuation of the solution by the simplex method.

We arbitrarily designate S_1 as the key row. The key number, then, is 20. Following the rules of transformation, we change the basic variables from S_1, S_2 (in Table 7.10) to Y, S_2 (in Table 7.11). The second program is shown in Table 7.11. Since all the $C_j - Z_j$ in Table 7.11 are either negative or zero, we have arrived at the optimal solution. However, the optimal solution is a *degenerate* basic feasible solution (the number of positive basic variables is less than m).

Table 7.11. Second and Optimal Tableau

Program (Basic Variables)	Profit per Unit C_b	Quantity	$C_j \rightarrow$ 40 X	50 Y	0 S_1	0 S_2
Y	50	250	1.25	1	0.05	0
S_2	0	0	-17.5	0	-1.5	1
Z_j			62.5	50	2.5	0
$C_j - Z_j$			-22.5	0	-2.5	0

In Figure 7.3 we present a graphical representation of our degenerate problem. The optimal solution of Table 7.11 corresponds to point A in Figure 7.3. It should be noted that in this problem the resolution of degeneracy was a simple matter. No matter which of the tied variables was removed first, we would have obtained the same solution, although the *number of iterations* would have been greater had we removed S_2 first. Two remarks must be made at this time. First, an arbitrary removal of one of the tied variables may mean that a much larger number of iterations will be necessary to arrive at the optimal solution than would be the case if some other tied variable were removed from the basis. Second, a more serious situation may arise if an arbitrary selection of the tied variable leads us to

---

* If we had denoted X, Y, S_1, S_2 by X_1, X_2, X_3, X_4, then, from Table 7.10, X_3 would be removed first, since it would have the smaller subscript of the tied variables. That is, S_1 would be removed.

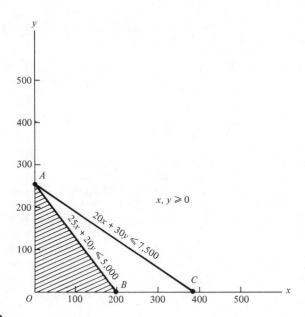

Figure 7.3

what we earlier called *cycling*. In cycling, we start from some basis and, after a few iterations, return to the same basis, so that an optimal solution may never be reached.

Although cycling is a theoretical possibility, it seldom occurs in practical problems. However, general methods of resolving degeneracy have been devised which, if followed, will ensure against falling into the cycle process.

7.8 SUMMARY

The solution of a linear programming problem by the simplex method rests on a simple procedure consisting essentially of three phases. The first phase, of course, is to formulate the problem and express it mathematically in terms of a linear objective function, linear structural constraints, and the non-negativity constraints. The second phase involves the design of an initial program that usually includes only the slack or the surplus variables.* The third phase consists in applying a test of optimality to determine whether a given program can be improved. The mechanics of the third phase, which repeats itself until an optimal solution (if it exists) is obtained, consists of two parts: (1) testing the optimality of the current program, and (2) revising the current program, if necessary, according to definite rules of transformation for the key row and the nonkey rows of the simplex tableau

---

*In minimization problems, some of the basic variables will be artificial variables. See Section 7.5.

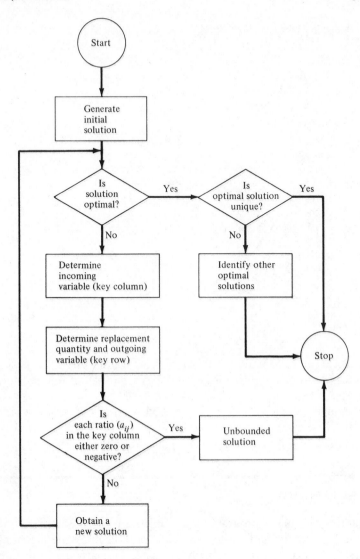

Figure 7.4   A Schematic of the Simplex Method

that contains the current program. A schematic diagram of this iterative procedure is given in Figure 7.4.

A step-by-step summary of the simplex procedure is given next.

STEP 1. Formulate the problem.

a. Make a precise statement of the objective function, and translate the technical specifications of the problem into a set of linear structural constraints (these are usually inequalities).

b. Convert the inequalities into equalities by the addition of slack or the subtraction of surplus variables. When needed, introduce artificial variables so that an initial basic feasible solution can easily be obtained.

c. Modify the objective function to include the slack, surplus, and/or the artificial variables along with their proper coefficients.*

STEP 2. Design an initial program (a basic feasible solution).

Design the first program so that only the slack and/or artificial variables are included in the solution. Place this program in a simplex tableau. In the C_j row, above each column variable, place the corresponding coefficient of that variable from step 1c.

STEP 3. Test for optimality and, if necessary, revise the program.

a. *Calculate the $C_j - Z_j$ numbers in the net evaluation row.* To get a number in the net evaluation row under a column, multiply the entries in that column by the corresponding profit (or cost) coefficients in the C_b column, and add the products (this gives us Z_j). Then subtract this sum from the number listed in the C_j row at the top of the column. Enter the resulting $C_j - Z_j$ number in the net evaluation row under the column.

b. *Test for optimality.* Examine the $C_j - Z_j$ entries in the net evaluation row for the given simplex tableau. If all the entries are zero or negative in maximization problems (zero or positive in minimization problems), the optimal solution has been obtained. Otherwise, the current program is not optimal and hence must be revised.

c. *Revise the program.*

1. *Find the key column.* The column under which the largest positive $C_j - Z_j$ falls is the key column in maximization problems (largest negative $C_j - Z_j$ identifies the key column in minimization problems).

2. *Find the key row and the key number.* Divide the entries in the "Quantity" column by the corresponding positive entries† of the key column to yield replacement quantities, and compare these replacement quantities. The row in which the smallest replacement quantity

---

* These coefficients are zero for all slack or surplus variables. In maximization problems, the coefficient of each artificial variable is $-M$; in minimization problems, it is $+M$.

† During any solution stage, the entries in the key column (substitution ratios) of a particular simplex tableau can present us with three situations. First, one of the positive entries may yield the smallest replacement quantity and definitely identify the key row. In this case, the simplex algorithm is applied in a straightforward manner. Second, the substitution ratios may be such that there is a "tie" between two or more minimum replacement quantities. In such a case, the problem becomes degenerate; however, the simplex method may be continued according to rules described in Section 7.7.2. Third, it may happen that all the entries in the key column are either zero or negative. In this case, the incoming variable can be introduced into the program without any limit, and no current basic variable can be removed from the solution. On the other hand, the values of the basic variables either remain the same or increase in magnitude without any limit (since all ratios of substitution are either zero or negative). Thus, in the last case, we have a situation in which the objective function is unbounded.

falls is the key row. The number that lies at the intersection of the key row and the key column is the key number.

3. *Transform the key row.* Divide all the numbers in the key row (starting with and to the right of the "Quantity" column) by the key number. The resulting numbers form the corresponding row of the next tableau.

4. *Transform the nonkey rows.* Subtract from the old row number of a given nonkey row (in each column) the product of the corresponding key row number and the corresponding fixed ratio formed by dividing the old row number in the key column by the key number. The result will give the corresponding new row number. Make this transformation for all the nonkey rows.

5. Enter the results of step 3c (transformed key and nonkey rows) in a tableau that represents the revised program.

STEP 4. Obtain the optimal program.

Repeat steps 3b and 3c until an optimal program has been derived.

Now that we know the procedure of the simplex algorithm for both the maximization and minimization problems, we present, in Chapter 8, the important concept of the dual problem in linear programming.

REFERENCES

Beale, E. M. L. *Applications of Mathematical Programming Techniques.* New York: American Elsevier Publishing Co., Inc., 1970.

Childress, R. L. *Sets, Matrices, and Linear Programming.* Englewood Cliffs, N.J.: Prentice-Hall, Inc., 1974.

Churchman, C. W., R. L. Ackoff, and E. L. Arnoff. *Introduction to Operations Research.* New York: John Wiley & Sons, Inc., 1957.

Cooper, L., and D. Steinberg. *Linear Programming.* Philadelphia: W. B. Saunders Company, 1974.

Daellenbach, H. G., and E. G. Bell. *User's Guide to Linear Programming.* Englewood Cliffs, N.J.: Prentice-Hall, Inc., 1970.

House, W. C. *Operations Research—An Introduction To Modern Applications.* Philadelphia: Auerbach Publishers, Inc., 1972.

Hughes, A. J., and D. Grawiog. *Linear Programming.* Reading, Mass.: Addison-Wesley Publishing Company, Inc., 1973.

Loomba, N. P. *Linear Programming.* New York: McGraw-Hill Book Company, 1964.

Loomba, N. P., and E. Turban. *Applied Programming for Management.* New York: Holt, Rinehart and Winston, Inc., 1974.

McMillan, C. *Mathematical Programming.* New York: John Wiley & Sons, Inc., 1974.

Simmons, D. M. *Linear Programming for Operations Research.* San Francisco: Holden-Day, Inc., 1972.

Spivey, W. A., and R. M. Thrall. *Linear Optimization.* New York: Holt, Rinehart and Winston, Inc., 1970.

Strum, J. E. *Introduction to Linear Programming.* San Francisco: Holden-Day, Inc., 1972.

REVIEW QUESTIONS AND PROBLEMS

7.1. Describe the rationale of the simplex method.

7.2. Why is the simplex method more efficient than the algebraic and the vector methods?

7.3. What is your understanding of the term "opportunity cost"? How is the concept of opportunity cost employed in designing a test of optimality?

7.4. What is the significance of $C_j - Z_j$ numbers?

7.5. A company makes impellers for pumps as one of its products. At present the firm makes two types of impellers, the standard type, which is good for general-purpose pumps, and the anticorrosion type, which is used for handling corrosive fluids. The standard type requires 8 units of carbon steel and 5 units of alloy steel per 100 impellers, whereas the anticorrosion type requires 3 units of carbon steel and 10 units of alloy steel per 100 impellers. The company has available to it during the next month 2,000 units of carbon steel and 1,000 units of alloy steel. How can the company best use its resources to maximize the profit on the impellers? Assume that a profit of $30 can be obtained on each 100 standard impellers and $40 on each 100 anticorrosion impellers.

7.6. A group of young people decided to leave the city to live on a farm. Toward that end, they purchased a 3-acre farm and decided to grow corn on a portion of the farm. There are two types of fertilizers available on the market. One unit of Rapidgrow fertilizer contains 8 pounds of ingredient A and 3 pounds of ingredients B, both of which are required for growing a good crop. One unit of Fastgrow brand contains 4 pounds of ingredient A and 9 pounds of ingredient B. A former agriculture technology student belonging to the group estimated that they will need 150 pounds of ingredient A and 100 pounds of ingredient B. At a cost of $8 and $10, respectively, for Rapidgrow and Fastgrow brands, what is the optimal quantity of each brand that should be purchased to meet the fertilizer requirements?

7.7. Maximize $\quad 30X_1 + 12X_2 + 18X_3 + 4X_4$

subject to the constraints

$$4X_1 + 2X_2 + 10X_3 + 1.2X_4 \leqslant 20$$
$$6X_1 + 2X_2 + 6X_3 + 0.5X_4 \leqslant 24$$
$$14X_1 \qquad\qquad + 2X_4 \leqslant 70$$

Assume the usual nonnegativity constraints.

7.8. A car manufacturer makes three kinds of small-car engines: model A, model B, and model C. Production of model A requires 2 hours of assembling, 1 hour of finishing, and 3 hours of testing. Production of model B requires 15 hours, 2 hours, and 2.5 hours of assembling, finishing, and testing, respectively. Production of model C requires 3 hours of assembling, 1.8 hours of finishing, and 4 hours of testing. The profits from model A, model B, and model C are $25, $35, $47.50, respectively. There are 120 hours of assembling, 95 hours of finishing, and 175 hours of testing available. How many engines of each model should be produced?

7.9. A mining company processes three kinds of ores A, B, and C through two sequential processes of wet grinding and drying. One unit of ore A takes 18 hours to grind and 12 hours to dry. One unit of ore B takes 12 hours to grind and 4 hours to dry, and one unit of ore C takes 2 hours to grind and 2 hours to dry. The profits obtained from a unit of ore A, ore B, and ore C are \$52, \$30, and \$9.50, respectively. How much of each type of ore should the company process to maximize the profits if 40 hours and 20 hours of grinding and drying time is available, respectively? How much is the profit?

7.10. Minimize
$$3X_1 + 5X_2 + 4X_3$$

subject to the constraints

$$
\begin{aligned}
6X_1 + 13X_2 &\geqslant 60 \\
11X_2 + 8X_3 &\geqslant 85 \\
2X_3 &\geqslant 25 \\
X_1, \quad X_2, \quad X_3 &\geqslant 0
\end{aligned}
$$

7.11. Minimize
$$15X_1 + 8X_2$$

subject to the constraints

$$
\begin{aligned}
2X_1 + 5X_2 &\leqslant 16 \\
4X_1 + 7X_2 &= 22 \\
10X_1 + 12X_2 &\geqslant 12 \\
X_1, \quad X_2 &\geqslant 0
\end{aligned}
$$

7.12. Minimize
$$2X_1 - X_2 + 4X_3$$

subject to the constraints

$$
\begin{aligned}
4X_1 - 3X_2 + 2X_3 &\geqslant 9 \\
- X_2 + 2X_3 &\geqslant 4 \\
4X_1 + X_2 + 8X_3 &\geqslant 21 \\
X_1, \quad X_2, \quad X_3 &\geqslant 0
\end{aligned}
$$

7.13. The manager of the production department of Lehigh Steel Company, Inc., has complete control of all functions within his department. He is also responsible for the scheduling of production.

All production is for storage only, with fabrication to follow after a slight delay. The production schedule is thus unaffected by the pattern of orders, and the production manager is free to schedule production so as to maximize profit.

For the current production period the company is concerned only with the production of nickel and chromium alloy steels. The production department receives 15 tons of nickel daily, along with a like amount of chromium. From these two alloying agents the company can produce alloy types 2010, 2020, and 1020. Alloy 2010 requires 200 pounds of nickel and 100 pounds of chromium per ton. Alloy 2020 requires 200 pounds of nickel and 200 pounds of chromium. Alloy 1020 requires 100 pounds of nickel and 200 pounds of chromium. The alloys net \$25, \$30, and \$35 profit per ton produced, respectively.

Assuming that 2 tons of alloy 2010 must be produced, design the most profitable production program.

7.14. West Chester, a small eastern city of 15,000 people, requires an average of 300,000 gallons of water daily. The city is supplied from a central waterworks, where the water is purified by such conventional methods as filtration and chlorination. In addition, two different chemical compounds, softening chemical and health chemical, are needed for softening the water and for health purposes. The water-works plans to purchase two popular brands that contain these chemicals. One unit of the Chemco Corporation's product gives 8 pounds of the softening chemical and 3 pounds of the health chemical. One unit of the American Chemical Company's product contains 4 pounds and 9 pounds per unit, respectively, for the same purposes.

To maintain the water at a minimum level of softness and to meet a minimum in health protection, experts have decided that 150 and 100 pounds of the two chemicals that make up each product must be added to the water daily. At a cost of $8 and $10 per unit, respectively, for Chemco's and American Chemical's products, what is the optimal quantity of each product that should be used to meet the minimum level of softness and a minimum health standard?

7.15. The H.E.E. Construction Company is building roads on the side of South Mountain. It is necessary to use explosives to blow up the boulders under the ground to make the surface level. There are three liquid ingredients (A, B, and C) in the liquid explosive used. It is known that at least 10 ounces of the explosive has to be used to get results. If more than 20 ounces is used, the explosion will be too damaging. Also, to have an explosion, at least $\frac{1}{4}$ ounce of ingredient C must be used for every ounce of ingredient A, and at least 1 ounce of ingredient B must be used for every ounce of ingredient C. The costs of ingredients A, B, and C are $6, $18, and $20 per ounce, respectively. Find the least-cost explosive mix necessary to produce a safe explosion.

7.16. The Roadhog Truck Company manufactures two truck types, the Snort and the Razorback. The Snort is a 50-ton truck, whereas the Razorback is 40 tons. The company has unlimited demand for both trucks, but plant facilities limit pro-duction. Each truck must pass through the three departments of the plant. The man-hours needed for each truck and the total man-hours available per month are given in Table 7.12. The profit from each Snort is $2,000, and the profit from each Razorback is $2,600. How should monthly production be scheduled to maximize profits?

Table 7.12

Department	Man-Hours Needed per Truck		Man-Hours Available per Month
	Snort	Razorback	
Motor	30	40	1,000
Assembly	20	11	275
Painting	4	5	335

7.17. The Wild Horses Oil Company makes three brands of gasoline: Man o' War, Trigger, and Swayback. Wild Horses makes its products by blending two grades of gasoline, each with a different octane rating. Each brand of gas must have an octane rating greater than a predetermined minimum. In gasoline blending, final octane rating is linearly proportional to component octanes (i.e., a blend of

50 percent 100-octane and 50 percent 200-octane gasoline is 150 octane). The other relevant data are given in Tables 7.13 and 7.14. There is no limit on the amount of each gasoline that may be sold. Determine weekly production to maximize profits.

Table 7.13

Blending Component	Octane	Cost per Gallon (cents)	Supply per Week (gallons)
A	200	10	20,000
B	130	8	10,000

Table 7.14

Brand	Minimum Octane	Sale Price per Gallon (cents)
Man o' War	180	24
Trigger	160	21
Swayback	140	12

7.18. Let us consider a production planning problem in which the finished products consist of two components A and B. One unit of the finished product consists of 2 units of component A and 3 units of component B. These components are produced by two departments that use the same raw materials but different methods of production. The components require two different raw materials that are available in limited quantities. The available supplies for the planning horizon are 300 units of raw material 1 and 400 units of raw material 2. The data for the problem are shown in Table 7.15. Design the optimal mix of the two production processes to maximize the output of the finished products.

Table 7.15

Department	Input per Production Run (units)		Output per Production Run (units)	
	Raw Material 1	Raw Material 2	Component A	Component B
I	3	8	6	4
II	6	4	5	6

8
The Dual and Sensitivity
Analysis

8.1 INTRODUCTION

Linear programming problems exist in pairs. Thus in linear programming, associated with every maximization problem is a minimization problem. Conversely, associated with every minimization problem is a maximization problem. As we shall see in this chapter, it is possible to state an original linear programming problem (maximization or minimization) and derive its dual problem according to well-known relationships of duality.* The original linear programming problem is called the *primal problem*, and the derived problem is called the *dual problem*.

The concept of the dual problem is important for several reasons.† We mention two here. First, the variables of the dual problem can convey important information to managers in terms of formulating their future plans. This idea was discussed in Section 3.6. Second, in some cases the dual problem can be instrumental in arriving at the optimal solution to the original problem in much less number of iterations (thereby reducing the cost and time of computations). For example, it takes much less computational effort to solve by the simplex method the dual of a linear programming problem whose original dimension is, say, $5,000 \times 2$.‡

* See Section 3.4 of Loomba and Turban [1974].
† See Loomba and Turban [1974, p. 91].
‡ The dual of this problem will have a dimension of $2 \times 5,000$. That is, an $m \times n$ primal problem reflects itself as an $n \times m$ dual problem.

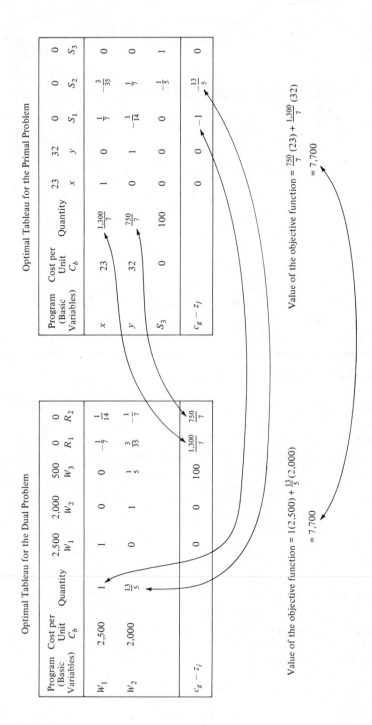

Figure 8.1 Comparison of the Primal and Dual Optimal Tableaus

The format of the simplex method is such that when the primal problem is solved, the dual is automatically being solved. Thus the optimal solution to the dual can be read from the optimal tableau of the primal problem (see Figure 8.1).

The concept of the dual will be explained in this chapter by means of an example.

8.2 THE DUAL TO A MAXIMIZATION PROBLEM

Let us consider the maximization problem of Table 8.1. As previously, the mathematical statement of the problem is

Maximize $23X + 32Y$

subject to the constraints

$$10X + 6Y \leq 2{,}500 \text{ (machine-hours)}$$
$$5X + 10Y \leq 2{,}000 \text{ (machine-hours)}$$
$$1X + 2Y \leq 500 \text{ (man-hours)}$$
$$X, \quad Y \geq 0$$

Table 8.1

Department	Product		Resource Capacity per Time Period
	A	B	
Cutting	10	6	2,500
Folding	5	10	2,000
Packaging	1	2	500
Profit per Unit	$23	$32	

The maximization problem stated above is our primal problem. Associated with it is a linear programming minimization problem that is dual to the given problem. We propose to formulate the dual by means of an intuitive argument.

Think of the primal problem here as the *seller's maximization problem* because the seller wishes to maximize his profits. Also, it should be noted that (1) the technology is fixed, and (2) the profits are generated because the seller has certain resources. Therefore, it is obvious that it will be useful, for purposes of managerial planning, to know what profits are generated by the respective resources. We answer this question by formulating a hypothetical buyer's problem.

Associated with the seller's maximization problem is a *buyer's minimization problem*. Let us explain the rationale. The buyer, it is assumed, will consider the purchase of the resources in full knowledge of the technical specifications given in Table 8.1.* If the buyer wishes to get an idea of his total outlay, he will have to determine how much he must pay to buy all the resources. Assume that he designates variables W_1, W_2, and W_3, to represent the per unit price or value that he will assign to cutting, folding, and packaging capacities, respectively, while making his purchase plans. The total outlay, which the buyer wishes to minimize, will be determined by the function $2,500W_1 + 2,000W_2 + 500W_3$. The objective function of the buyer, therefore, is

Minimize $\qquad\qquad 2,500W_1 + 2,000W_2 + 500W_3$ $\qquad\qquad$ (8.1)

The linear function of (8.1) must be minimized in view of the knowledge that the current technology yields a profit of $23 by spending 10 machine-hours of the cutting department, 5 machine-hours of the folding department, and 1 man-hour of the packaging department. Hence the specific values for the variables W_1, W_2, and W_3 must be assigned in such a manner that $10W_1 + 5W_2 + 1W_3$ generates *at least* $23. By the same reasoning, and *simultaneously*, these assigned values must be such that $6W_1 + 10W_2 + 2W_3$ generates *at least* $32. Hence, in the buyer's problem, the two structural constraints are

$$10W_1 + 5W_2 + 1W_3 \geqslant 23$$
$$6W_1 + 10W_2 + 2W_3 \geqslant 32$$

In addition, the assigned values to the resources in our problem must be nonnegative. That is, W_1, W_2, and W_3 must each be greater than or equal to zero.

We can now state the dual to the given primal problem.

Minimize $\qquad\qquad 2,500W_1 + 2,000W_2 + 500W_3$

subject to the constraints

$$10W_1 + 5W_2 + 1W_3 \geqslant 23$$
$$6W_1 + 10W_2 + 2W_3 \geqslant 32$$
$$W_1, \qquad W_2, \qquad W_3 \geqslant 0$$

Let us make some observations regarding the dual problem stated above.

1. The variables W_1, W_2, W_3, etc., are called the *dual variables*. The values assigned to the dual variables in the optimal tableau of the dual problem represent artificial accounting prices, or implicit prices, or shadow prices, or marginal worths of the various resources. As mentioned in Section 8.4, the values of the dual variables can also be read from the net evaluation row of the optimal tableau of the primal problem. That is, the $C_j - Z_j$ of the

---

* This is because linear programming is a deterministic model that assumes static conditions and full knowledge of all actions as well as their consequences. See Section 2.3.

optimal simplex tableau of the primal automatically yield the optimal solution to the dual. (See Figure 8.1.)

The dimension of any dual variable is determined by the units of the constraint to which the dual variable corresponds. Thus note that the dimension of W_1 as well as W_2 is dollars/machine-hours, while the dimension of W_3 is dollars/man-hours. Similarly, if a constraint had been specified in, say, pounds, the dimension of its dual variable would have been dollars/pound. Furthermore, if one more (marginal) unit of a specified resource is made available, the value of the objective function will increase by an amount equal to the value of the dual variable associated with that resource.

2. Since, by definition, the entire profit in the maximization must be traced to the given resources, the buyer's total outlay, at the equilibrium point, must equal the total profit. That is, the optimal value of the objective function of the primal equals the optimal value of the objective function of the dual.

3. The example that we have provided is that of a *symmetrical dual*. Here all structural constraints are inequalities, and all variables are restricted to nonnegative values.* The "symmetry" in our example can be observed if we place the primal and the dual next to each other.

Primal	Dual
Maximize $23X + 32Y$	Minimize $2,500W_1 + 2,000W_2 + 500W_3$
subject to the constraints	subject to the constraints
$10X + 6Y \leqslant 2,500$	$10W_1 + 5W_2 + 1W_3 \geqslant 23$
$5X + 10Y \leqslant 2,000$	$6W_1 + 10W_2 + 2W_3 \geqslant 32$
$1X + 2Y \leqslant 500$	$W_1, \quad W_2, \quad W_3 \geqslant 0$
$X, \quad Y \geqslant 0$	

Note that:

1. If in the primal, the objective function is to be maximized; then in the dual it is to be minimized. Conversely, if in the primal the objective function is to be minimized, then in the dual, it is to be maximized.
2. Objective function coefficients of the primal appear as right-hand-side requirements in the dual, and vice versa.
3. The right-hand-side numbers of the primal appear as objective function coefficients in the dual, and vice versa.
4. The input–output coefficient matrix of the dual is the transpose of the input–output coefficient matrix of the primal, and vice versa.

* For a discussion of unsymmetrical dual, see Loomba and Turban [1974, p. 92].

5. If the inequalities in the primal are of the "less than or equal to" type; then in the dual they are of the "greater than or equal to" type. Conversely, if the inequalities in the primal are of the "greater than or equal to" type; then in the dual, they are of the "less than or equal to" type.

These comments and some additional items of correspondence are now summarized in a table.*

Primal	Dual
Maximize	Minimize
Objective function	Right-hand side
Right-hand side	Objective function
ith row of input–output coefficients	ith column of input–output coefficients
jth column of input–output coefficients	jth row of input–output coefficients
ith relation an inequality ($\leq$)	ith variable nonnegative
ith relation an equality ($=$)	ith variable unrestricted in sign
jth variable nonnegative	jth relation an inequality ($\geq$)
jth variable unrestricted in sign	jth relation an equality ($=$)

8.3 SOLVING THE DUAL PROBLEM

The simplex algorithm can now be applied to the following dual problem:

Minimize

$$2{,}500W_1 + 2{,}000W_2 + 500W_3 - 0R_1 - 0R_2 + MA_1 + MA_2$$

subject to the constraints

$$10W_1 + \quad 5W_2 + \quad 1W_3 - 1R_1 + 0R_2 + \quad 1A_1 + \quad 0A_2 = 23$$
$$6W_1 + \quad 10W_2 + \quad 2W_3 + 0R_1 - 1R_2 + \quad 0A_1 + \quad 1A_2 = 32$$

and

$$W_1, \qquad W_2, \qquad W_3, \quad R_1, \quad R_2, \quad A_1, \quad A_2 \geq 0$$

The various solution stages are summarized in Tables 8.2 to 8.4. Since all the $C_j - Z_j$ are positive or zero in Table 8.4, the third tableau represents the optimal program. The optimal program is

$$W_1 = 1, W_2 = 2.6 \qquad \text{and} \qquad W_3 = 0, R_1 = 0, R_2 = 0$$

* See Loomba and Turban [1974, p. 92].

Table 8.2. First Simplex Tableau

Program (Basic Variables)	Cost per Unit C_b	Quantity	$C_j \to$ 2,500 W_1	2,000 W_2	500 W_3	0 R_1	0 R_2	M A_1	M A_2	Replacement Quantity
A_1	M	23	10	5	1	-1	0	1	0	$2\frac{3}{10} \to$ outgoing variable
A_2	M	32	6	10	2	0	-1	0	1	$5\frac{1}{3}$
Z_j			$16M$	$15M$	$3M$	$-M$	$-M$	M	M	
Net evaluation row $(C_j - Z_j)$			$2,500 - 16M$	$2,000 - 15M$	$500 - 3M$	M	M	0	0	

incoming variable

Table 8.3. Second Simplex Tableau

Program (Basic Variables)	Cost per Unit C_b	Quantity	$C_j \to$ 2,500 W_1	2,000 W_2	500 W_3	0 R_1	0 R_2	M A_1	M A_2	Replacement Quantity
W_1	2,500	2.3	1	0.5	0.1	-0.1	0	0.1	0	4.6
A_2	M	18.2	0	7	1.4	0.6	-1	-0.6	1	2.6 → outgoing variable
Z_j		2,500	2,500	$1{,}250 + 7M$	$250 + \frac{14}{10}M$	$-250 + \frac{6}{10}M$	$-M$	$250 - \frac{6}{10}M$	M	
$C_j - Z_j$			0	$750 - 7M$	$250 - 1.4M$	$250 - 0.6M$	M	$-250 + 1.6M$	0	

incoming variable (W_2)

Table 8.4. Third and Optimal Tableau

Program (Basic Variables)	Cost per Unit C_b	Quantity	$C_j \to$ 2,500 W_1	2,000 W_2	500 W_3	0 R_1	0 R_2	M A_1	M A_2
W_1	2,500	1	1	0	0	$-\frac{1}{7}$	$\frac{1}{14}$	$\frac{1}{7}$	$-\frac{1}{14}$
W_2	2,000	2.6	0	1	0.2	$\frac{3}{35}$	$-\frac{1}{7}$	$-\frac{3}{35}$	$\frac{1}{7}$
Z_j			2,500	2,000	400	$-\frac{1,300}{7}$	$-\frac{750}{7}$	$\frac{1,300}{7}$	$\frac{750}{7}$
$C_j - Z_j$			0	0	100	$\frac{1,300}{7}$	$\frac{750}{7}$	$M - \frac{1,300}{7}$	$M - \frac{750}{7}$

The meaning of this program is as follows:

W_1 = marginal worth of 1 unit of the cutting-department resource
  = \$1.00/machine-hour

W_2 = marginal worth of 1 unit of the folding-department resource
  = \$2.60/machine-hour

W_3 = marginal worth of 1 unit of the packaging-department resource
  = \$0.00/man-hour

When we examine the $C_j - Z_j$ numbers in the net evaluation row of Table 7.5 (optimal tableau for the primal), we note that the absolute values of the $C_j - Z_j$ under the slack variables S_1, S_2, and S_3 are precisely 1, 2.6, and 0. The fact that the marginal value of the packaging-department resource is zero is supported by Table 7.5. Note that the optimal program in Table 7.5 leaves 100 units of the packaging capacity unused.

In Figure 8.1 we compare the optimal tableaus of the primal and the dual and indicate their relationships to each other. Note that the A_1 and A_2 columns of Table 8.4 (optimal tableau) have been left out in Figure 8.1. This is perfectly all right because, as the reader will recall, the artificial variables have no physical interpretation. They are a device to obtain, in an easy manner, the first basic feasible solution, when all the constraints are either equalities or inequalities of the "greater than or equal to" type.

8.4 COMPARISON OF THE OPTIMAL TABLEAUS OF THE PRIMAL AND ITS DUAL

We place in Figure 8.1 the optimal tableau of the primal (Table 7.5) next to the optimal tableau of its dual (Table 8.4) and make some important observations. The objective functions of the two optimal tableaus assume identical values. We note further that $C_j - Z_j$ entries (with signs changed) in the net evaluation row under columns S_1 and S_2 of the optimal tableau of the primal are the same as entries under the "Quantity" column in the optimal dual tableau of the problem. That is, the optimal solution to the dual can be retrieved from appropriate $C_j - Z_j$ of the optimal simplex tableau of the primal. In addition, since S_3 is a basic variable in the optimal tableau of the primal, it indicates unused packaging capacity. Hence $W_3 = 0$, as can be inferred from the optimal tableau of the dual. Also, the magnitudes of the variables X and Y in the primal optimal tableau are exactly the same as the entries in the net evaluation row under columns R_1 and R_2 of the optimal tableau of the dual problem. This type of correspondence between the optimal tableaus of the primal and its dual always exists. Thus the solution to a primal problem in linear programming can always provide a solution to its dual. The relationships just described should be carefully checked and grasped by the reader.

8.5 SENSITIVITY ANALYSIS

Optimal solutions to linear programming problems are obtained under a set of assumptions such as fixed technology, fixed prices, and fixed levels of resources or requirements. These assumptions, implying certainty, complete knowledge, and static conditions permit us to design an optimal program. The conditions in the real world, however, might be different from those that are assumed by the model. It is, therefore, desirable to determine how sensitive the optimal solution is to different types of changes in the problem data and parameters. The changes whose effect on the optimal solution needs to be analyzed include (1) changes in such parameters as objective function coefficients (C_j), input–output coefficients (a_{ij}), resource or requirement levels (b_i); and (2) possible addition or deletion of products or methods of production. *Sensitivity analysis, optimality analysis,* and *parametric programming* are various names for investigating the relationships between optimal solutions and possible changes in various components of the problem.

Managers conduct sensitivity analysis when, in their planning processes, they ask what are called the "what-if" questions. The what-if questions are like a double-edged sword. They are designed to project the consequences of possible changes in the future, as well as the impact of the possible errors of estimation of the past. Thus the need for sensitivity analysis (or what-if analysis) arises from two sources. First, we want to know the effect of, and hence be prepared for, possible future changes in various parameters and components of the problem. Second, we want to know the degree of error in estimating certain parameters that could be absorbed by the current optimal solution. In other words, the sensitivity analysis answers questions regarding what errors of estimation could have been committed, or what possible future changes can occur, without disturbing the optimality of the current optimal solution.

The results of the sensitivity analysis establish *ranges* (upper limits and lower limits) for different parameters (C_j, a_{ij}, b_i, etc.) within which the current optimal program will remain optimal. In this sense, sensitivity analysis is a major guide to managerial planning and control. Further, it should be noted that sensitivity analysis avoids the need for *reworking the entire problem* from the very beginning each time a change is investigated or incorporated. Rather, the current optimal solution can be used to study the effect of changes with minimal computational effort. It is possible to illustrate sensitivity analysis with respect to changes in C_j, a_{ij}, and b_i; adding or deleting a new column (i.e., a new product); adding or deleting a new row (i.e., a new process). However, we shall restrict our presentation to investigating the effect of changes in the objective function (C_j) and changes in the resources or requirements (b_i).*

* For a more detailed analysis, see Loomba and Turban [1974, pp. 137–47].

8.5.1 Changes in the Objective Function Coefficients

The sensitivity of the optimal solution to changes in the objective function coefficients (C_j) is investigated by checking the new or changed coefficient, say, $C_j + G$, against the $C_j - Z_j$ numbers of the optimal solution. The variable G reflects the amount of change, and it can be positive or negative. For maximization problems, the current basis remains optimal as long as $(C_j + G) - Z_j$ remains nonpositive. Conversely, for minimization problems, the current basis remains optimal as long as $(C_j + G) - Z_j$ remains nonnegative.

Before illustrating sensitivity analysis for C_j, we advance some intuitive arguments to establish the required actions in carrying out such an analysis. Consider first the idea of a *nonbasic variable* in maximization problems. It is clear that a nonbasic variable can enter the solution only if its profit coefficient is increased beyond a certain limit. Hence, in maximization problems, we must determine the *upper limit* of the profit coefficient of each nonbasic variable. Beyond that limit the optimality of the current solution is destroyed. Conversely, in minimization problems, sensitivity analysis requires the determination of the *lower limit* of C_j for a nonbasic variable. Hence for nonbasic variables we have to determine either the upper limit (in maximization problems) or the lower limit (in minimization problems).

However, when it comes to conducting sensitivity analysis on the objective function coefficients of the *basic variables*, we must investigate the *range* (i.e., both the upper and lower limits). The determination of the range is necessary because an increase in the objective function coefficient of a basic variable would mean that, at some point, resources from other products should be diverted to this more profitable product. Similarly, a decrease in the objective function coefficient of the same basic variable would mean that, at some point, resources should be diverted away from this less-profitable product. We summarize the implications of our intuitive arguments in Table 8.5.

Table 8.5. Sensitivity Analysis for C_j

Type of Variable	Maximization Case	Minimization Case
Nonbasic	Determine the *upper limit*	Determine the *lower limit*
Basic	Determine the *range*	Determine the *range*

While determining the range (i.e., both the upper and lower limits) for C_j of a basic variable, it can happen that only one of the limits is meaningful. If such is indeed the case in a specific situation, it will be identified by the mechanics of the analysis, as we shall illustrate in the following example.

EXAMPLE 8.1

Let us illustrate sensitivity analysis for C_j with respect to the maximization problem whose technical specifications appear in Table 8.6. The optimal simplex tableau for the problem is given in Table 8.7. In the optimal solution we observe that X, S_2, and S_3 are *basic* variables, and Y and S_1 are *nonbasic* variables.

Table 8.6

Department	Product A	Product B	Capacity
Cutting	10	6	2,500
Folding	5	10	2,000
Packaging	1	2	500
Profit per Unit	$50	$25	

Table 8.7   Optimal Solution

Program (Basic Variables)	Profit per Unit C_b	Quantity	$C_j \rightarrow$ 50 X	25 Y	0 S_1	0 S_2	0 S_3
X	50	250	1	$\frac{3}{5}$	$\frac{1}{10}$	0	0
S_2	0	750	0	7	$-\frac{1}{2}$	1	0
S_3	0	250	0	0	$\frac{7}{5}$	0	1
Z_j			50	30	5	0	0
$C_j - Z_j$			0	-5	-5	0	0

Let us first consider the case of a *nonbasic* variable. In the optimal tableau of a maximization problem, a nonbasic variable (the product is not being produced) has a negative $C_j - Z_j$.* The negative sign indicates that the production of the product would, at that solution stage, decrease the profit. The question that we want to answer is this: By what amount will the profit coefficient of the nonbasic variable have to be increased to destroy the optimality of the current program? Let this amount be G_2 for the nonbasic variable Y, and let the corresponding net-profit coefficient be C'_2. That is, $C'_2 = C_2 + G_2$. The optimality of the solution shown in Table 8.7 is destroyed as soon as

$$C'_2 - Z_2 > 0 \quad \text{or} \quad (C_2 + G_2) - Z_2 > 0 \tag{8.2}$$

* In the case of multiple optimal solution, the $C_j - Z_j$ could be zero.

We see from column Y of Table 8.7 that $C_2 = 25$ and $Z_2 = 30$. Substituting these values in (8.2), we obtain $(25 + G_2) - 30 > 0$; or $G_2 > 5$. That is, the *change* in profit coefficient of the nonbasic variable Y will have to be more than 5 in order to destroy the optimality of the current optimal solution. Hence the upper limit for C_2 is 30. That is, as soon as the profit coefficient of Y is 30 or more, we must make Y a basic variable.

In our example, the only other nonbasic variable is S_1. Since S_1 is a slack variable, we shall not bother to conduct sensitivity analysis on its profit coefficient.

We repeat that in maximization problems we are interested in determining the upper limit beyond which the profit coefficient of a nonbasic variable will destroy optimality of the current solution. This upper limit for the profit coefficients of the nonbasic variables is determined from the criterion

$$(C_j + G) - Z_j > 0 \tag{8.3}$$

In minimization problems, we are interested in determining the lower limit below which the cost coefficient of a nonbasic variable will destroy optimality of the current solution. This lower limit for the cost coefficients of the nonbasic variables is determined from the criterion

$$(C_j + G) - Z_j < 0 \tag{8.4}$$

Let us now investigate the effect of the changes in a *basic* variable. In Table 8.7, we see that variables X, S_2, and S_3 are the basic variables. Let us consider variable X whose current profit coefficient is 50. Let us change this coefficient to $50 + G_1$ and then proceed with our analysis. The change in the profit coefficient of X is reflected in Table 8.8. Note that Table 8.8 is the same as Table 8.7 except that the profit coefficient of X has been changed from 50 to $50 + G_1$.

Table 8.8

Program (Basic Variables)	Profit per Unit C_b	Quantity	$C_j \rightarrow$ $50 + G_1$ X	25 Y	0 S_1	0 S_2	0 S_3
X	$50 + G_1$	250	1	$\frac{3}{5}$	$\frac{1}{10}$	0	0
S_2	0	750	0	7	$-\frac{1}{2}$	1	0
S_3	0	250	0	0	$\frac{7}{5}$	0	1
Z_j			$50 + G_1$	$30 + \frac{3}{5}G_1$	$5 + \frac{1}{10}G_1$	0	0
$C_j - Z_j$			0	$-5 - \frac{3}{5}G_1$	$-5 - \frac{1}{10}G_1$	0	0

Since this is a maximization problem, the solution shown in Table 8.8 will remain optimal only if each $C_j - Z_j \leqslant 0$. Hence, in order to determine the value of G_1, we must solve the following system of linear inequalities:

$$-5 - \tfrac{3}{5} G_1 \leqslant 0 \tag{8.5}$$

$$-5 - \tfrac{1}{10}G_1 \leqslant 0 \tag{8.6}$$

From the first inequality, $G_1 \geqslant -\tfrac{25}{3}$; from the second equality, $G_1 \geqslant -50$. Hence we conclude that the profit coefficient of X must decrease by more than $\tfrac{25}{3}$ in order to destroy the current optimal solution.

Note that we have established the lower limit* for the profit coefficient of X. The nature of the two inequalities (8.5) and (8.6) was such that only one side of the range was determined. If both sides of the range (i.e., the lower as well as the upper limit) were meaningful, the inequalities, such as (8.5) and (8.6), will generate both limits.

An examination of Table 8.8 indicates why an upper limit for the profit coefficient of X (i.e., C_1) is not meaningful in terms of sensitivity analysis. No matter how much C_1 is increased, it will never affect the optimality of the solution shown in Table 8.8. The product X is the only *real* variable in the current optimal solution.

Whenever a range exists for C_j of basic variables, it will automatically be established by the simultaneous solution of the linear inequalities shown in (8.7) and (8.8).

For maximization problems:

$$(C_j + G) - Z_j \leqslant 0 \qquad j = 1, 2, \ldots \tag{8.7}$$

For minimization problems:

$$(C_j + G) - Z_j \geqslant 0 \qquad j = 1, 2, \ldots \tag{8.8}$$

It should be noted that if sensitivity analysis were performed on the C_j's of the problem stated in Table 3.2 (the same problem as shown in Table 8.6 except that $C_1 = \$23$ and $C_2 = \$32$), then the two ranges for the profit coefficients will be given by:

	Lower Limit	Upper Limit
For C_1	16.00	$53\tfrac{1}{3}$
For C_2	13.80	46.00

The reader should verify these results.

---

* The lower limit for C_1, the profit coefficient of X, is $50 - \tfrac{25}{3} = 41\tfrac{2}{3}$. While conducting sensitivity analysis on a basic variable, it should be noted that, in the established range, the basic variables remain the same. But the *values* of the basic variables and objective function change.

8.5.2 Changes in the Resources or Requirements

Knowledge of how sensitive the optimal solution is to changes in resources or requirements (b_i) is often important to managers because of changing conditions in the marketplace. It is extremely useful for planning purposes to know the ranges within which each available resource, or stated requirement, can vary without affecting the feasibility of the current basic variables. It is interesting to note that changes in b_i affect the magnitudes of the basic variables as well as the value of the objective function. However, the $C_j - Z_j$ (the net evaluation row numbers) are not affected. Hence the current solution remains optimal until one of the basic variables becomes infeasible. It is this fact that we shall use to determine the range for each b_i.

It will be recalled from Section 5.13 that a system of linear equations can be represented in matrix notation in the form

$$\mathbf{AX} = \mathbf{b} \tag{8.9}$$

Assume that $\mathbf{A}$ is a square matrix. Then the system (8.9) can be solved by taking the inverse of $\mathbf{A}$ and multiplying both sides of Equation (8.9) by $\mathbf{A}^{-1}$:*

$$\mathbf{A}^{-1}[\mathbf{AX}] = \mathbf{A}^{-1}\mathbf{b}$$

or

$$\mathbf{X} = \mathbf{A}^{-1}\mathbf{b} \tag{8.10}$$

What Equation (8.10) says is this: If a solution vector $\mathbf{X}$ to a system of linear inequalities has been found, it implies the existence of an inverse matrix. In each simplex tableau, we have a solution vector $\mathbf{X}$ that is comprised of the basic variables. Hence each simplex tableau implies the existence of a basis† $\mathbf{B}$, and its inverse $\mathbf{B}^{-1}$ also exists. As a matter of fact, it is possible to retrieve the inverse of a basis from the body of the simplex tableau. For example, the basic variables in Table 7.5 are X, Y, and S_3. Hence the basis $\mathbf{B}$ for Table 7.5 consists of the vectors associated with basic variables X, Y, and S_3. That is,

$$\mathbf{B} = \begin{bmatrix} 10 & 6 & 0 \\ 5 & 10 & 0 \\ 1 & 2 & 1 \end{bmatrix}$$

and $\mathbf{B}^{-1}$ can be retrieved from Table 7.5 as follows:

$$\mathbf{B}^{-1} = \begin{bmatrix} \frac{1}{7} & -\frac{3}{35} & 0 \\ -\frac{1}{14} & \frac{1}{7} & 0 \\ 0 & -\frac{1}{5} & 1 \end{bmatrix}$$

---

* It will be recalled from Chapter 5 that $\mathbf{A}^{-1}\mathbf{A} = \mathbf{I}$ (the identity matrix); and $\mathbf{IX} = \mathbf{X}$.
† It will be recalled from Section 5.15 that a *basis* is a set of basic vectors.

We are now in a position to establish the mechanism for conducting sensitivity analysis on b_i. We shall utilize the problem of Table 3.2 to establish three steps needed for sensitivity analysis. The problem follows, and its optimal solution is reproduced as Table 8.9.

Table 8.9. Optimal Tableau

Program (Basic Variables)	Profit per Unit C_b	Quantity	$C_j \rightarrow$ 23 X	32 Y	0 S_1	0 S_2	0 S_3
X	23	$\frac{1,300}{7}$	1	0	$\frac{1}{7}$	$-\frac{3}{35}$	0
Y	32	$\frac{750}{7}$	0	1	$-\frac{1}{14}$	$\frac{1}{7}$	0
S_3	0	100	0	0	0	$-\frac{1}{5}$	1
Z_j			23	32	1	$\frac{13}{5}$	0
$C_j - Z_j$			0	0	-1	$-\frac{13}{5}$	0

EXAMPLE 8.2

Maximize $\qquad 23X + 32Y$

subject to the constraints

$$10X + 6Y \leqslant 2{,}500$$
$$5X + 10Y \leqslant 2{,}000$$
$$1X + 2Y \leqslant 500$$

and

$$X, \quad Y \geqslant 0$$

The following steps are now needed.

STEP 1. Determine, from the statement of the original problem, the right-hand-side resource (or requirement) vector.

In our case, the requirement vector $\mathbf{b}$ is

$$\mathbf{b} = \begin{bmatrix} 2{,}500 \\ 2{,}000 \\ 500 \end{bmatrix}$$

STEP 2. Add a variable G_i to that b_i for which the range of "acceptable" variation is being established.

Let us investigate the range for the first resource. The new capacity level for the first resource is $2{,}500 + G_1$.

STEP 3. The range (the upper and lower bounds) for any b_i is then determined from the requirement that the current basic variables remain feasible (i.e., $\mathbf{X}_b = \mathbf{B}^{-1}\mathbf{b} \geqslant \mathbf{0}$, where $\mathbf{X}_b$ refers to the vector of basic variables).

In our example, as discussed previously,

$$\mathbf{B}^{-1} = \begin{bmatrix} \frac{1}{7} & -\frac{3}{35} & 0 \\ -\frac{1}{14} & \frac{1}{7} & 0 \\ 0 & -\frac{1}{7} & 1 \end{bmatrix} \quad \text{and the changed} \quad \mathbf{b} = \begin{bmatrix} 2{,}500 + G_1 \\ 2{,}000 \\ 500 \end{bmatrix}$$

Hence $\mathbf{B}^{-1}\mathbf{b} \geqslant \mathbf{0}$ implies that

$$\begin{bmatrix} \frac{1}{7} & -\frac{3}{35} & 0 \\ -\frac{1}{14} & \frac{1}{7} & 0 \\ 0 & -\frac{1}{5} & 1 \end{bmatrix} \begin{bmatrix} 2{,}500 + G_1 \\ 2{,}000 \\ 500 \end{bmatrix} \geqslant \begin{bmatrix} 0 \\ 0 \\ 0 \end{bmatrix} \tag{8.11}$$

We obtain the following inequalities from (8.11),

$$G_1 \geqslant -1{,}300$$
$$G_1 \leqslant \ \ \ 1{,}500$$

This means that for $-1{,}300 \leqslant G_1 \leqslant 1{,}500$, the current basic variables X, Y, and S_3 remain in the solution. That is, a range of 1,200 to 4,000 is permitted for b_1.* In this range, the basic variables remain the same (that is, the basis is unchanged), but the *values* of the basic variables and objective function change. The values of the dual variables are *not* altered.

By following a similar analysis we can determine the following ranges for b_2 and b_3: for b_2, the range is $1{,}250 \leqslant b_2 \leqslant 2{,}500$; for b_3, the range is $400 \leqslant b_3 \leqslant \infty$.

8.6 SUMMARY

In this chapter we have presented the important concept of the dual problem in linear programming. Linear programming problems exist in pairs, the primal and its dual. Relationships between the primal and the dual were discussed and their correspondence was established. The derivation of the dual problem, by utilizing the primal, was illustrated. We then presented the idea of sensitivity analysis and discussed how managers can use sensitivity analysis for purposes of planning and control. How to conduct sensitivity analysis on objective function coefficients (C_j) and the right-hand-side resource or requirement levels (b_i) was illustrated with specific examples. The concept of the dual variable is important for several reasons. For example, it is helpful in assigning marginal worth to a given resource so that

---

* The quantity $2{,}500 + G_1$ yields the range as $(2{,}500 - 1{,}300)$ to $(2{,}500 + 1{,}500)$.

rational decisions can be made in terms of acquiring additional resources from the marketplace. The concept of the dual variable has also been used in designing a method for testing the optimality of transportation programs. In Chapter 9, we present the transportation model and discuss how dual variables can be utilized in testing and designing optimal transportation programs.

REFERENCES

See references at the end of Chapter 7.

REVIEW QUESTIONS AND PROBLEMS

8.1. What is the significance of duality? Give an economic interpretation to the values of the dual variables.

8.2. What managerial insights or benefits can be gained by conducting a sensitivity analysis? Is sensitivity analysis limited to linear programming?

8.3. Formulate the dual of the following linear programming problem?

(a) Maximize $\qquad 10X_1 + 19X_2$

subject to the constraints

$$5X_1 + 6X_2 \leqslant 20$$
$$7X_1 + 13X_2 \leqslant 39$$
$$-4X_1 + 8X_2 \leqslant 8$$

and

$$X_1, \quad X_2 \geqslant 0$$

(b) What is the interpretation of the dual variables?

8.4. Maximize $\qquad 23X + 32Y + 18Z$

subject to the constraints

$$10X + 6Y + 2Z \leqslant 2,500$$
$$5X + 10Y + 5Z \leqslant 2,000$$
$$X + 2Y + 2Z \leqslant 500$$

and

$$X, \quad Y, \quad Z \geqslant 0$$

(a) Write and solve the dual of this problem.

(b) Will the dual optimal solution remain optimal if the profit coefficient of X is reduced from 23 to 15?

(c) Will the dual optimal solution change if the resource capacity of the first constraint in the primal problem is increased to 3,000?

8.5. The management of American Oil Company has received orders from International Airlines to supply 500 units of gasoline A and 750 units of gasoline

B. Amoco has two processes that can be used for the manufacture of gasolines *A* and *B*. One hour of process I produces 5 units of gasoline *A* and 10 units of gasoline *B*. One hour of process II produces 8 units of gasoline *A* and 6 units of gasoline *B*. One hour of process I uses a particular combination of crudes that cost $7, and 1 hour of process II uses a different combination of crudes, which costs $9.20.

(a) Find the optimal number of hours for which process I and process II should be run.

(b) Will the optimal solution change if the cost of 1 hour of process I rises to $15?

8.6. (a) Design an optimal program for Problem 3.4.

(b) Will the optimal solution change if the capacity of the cutting section is enlarged to 55 hours and that of sanding section reduced to 45?

8.7. The alloying metals of iron, nickel, and chrome are to be mixed to form an alloy to be used in the manufacture of valve seats. The proportion of iron has to be limited to 4 times the proportion of nickel. The sum of proportions of nickel and chrome has to be at least 4 per cent of the proportion of iron. The alloying metals have an impurity content of 0.1 per cent, 0.03 per cent, and 0.2 per cent, respectively. It is known from the laboratory tests that the percentage of sulfur impurity in the alloy should not exceed 0.075 per cent. If the metals cost $1, $5, and $7 per kilogram, respectively:

(a) Find the optimal proportions of the alloying metals.

(b) What is the dual to this problem? How will you interpret the dual variables?

(c) Will the optimal solution change if the cost of nickel increases by 20 per cent?

8.8. A small winery manufactures two types of wine, Burbo's Better and Burbo's Best. Burbo's Better sells for $20 per quart, whereas Burbo's Best sells for $30 per quart. Two men mix the two wines. It takes a man 2 hours to mix a quart of Burbo's Better and 6 hours to mix a quart of Burbo's Best. Each man works an 8-hour day, with 30 minutes for lunch. The quantity of alcohol that the plant can use is limited to 18 ounces daily. Six ounces are used in each quart of Burbo's Better, and 3 ounces in each quart of Burbo's Best. How many quarts of each should the winery make? Solve this problem via its dual.

8.9. Wung Lee's Chinese Restaurant serves two types of chow mein, chicken and shrimp. There is a $2 profit on each pound of shrimp chow mein, and a $1 profit on chicken chow mein. From past experience Wung Lee knows that he should not make more than 10 pounds of chicken chow mein and 8 pounds of shrimp chow mein each day. The two cooks work an 8-hour day, with 30 minutes for lunch. A pound of chicken chow mein takes 2 hours to cook, whereas a pound of shrimp chow mein takes 1 hour. How many pounds of each type of chow mein should Wung Lee make? Utilize the dual approach in solving this problem.

8.10. The Catchem Every Time Company manufactures two types of baseball gloves. The profit from the high-quality glove is $4, and the profit from the low-quality glove is $2. Machine *A* is the only machine on which these gloves can be made. It takes 30 minutes to manufacture a low-quality glove, whereas 2 hours are needed to make a high-quality glove. The plant works two 8-hour shifts. There is enough material to make 24 low-quality gloves per day. It takes twice as much material to manufacture a high-quality glove. How many of each type of glove should the company make? Form the dual and obtain the values of the dual variables from a regular simplex solution of this problem.

9

The Transportation Model

9.1 INTRODUCTION

The transportation model deals with a special class of linear programming problems in which the objective is to "transport" a *homogeneous* commodity from various "origins" to different "destinations" at a minimum total cost.* The supply available at each origin and the quantity demanded by each destination are given in the statement of the problem. Also given is the cost of shipping a unit of goods from a known origin to a known destination. As in the linear programming problems discussed in previous chapters, all relationships are assumed to be linear.

Given the information regarding the total capacities of the origins, the total requirements of the destinations, and the shipping cost per unit of goods for available shipping routes, the transportation model is used to determine the optimal shipping program(s) that results in minimum total shipping costs. The transportation model can be extended to solve problems related to topics such as production planning, machine assignment, and plant location.

Insofar as the transportation problem is a special case of the general linear programming problem, it can always be solved by the simplex method. However, the transportation algorithm, which we shall develop in later sections of this chapter, provides a much more efficient method of handling

* Of course, if the payoff measure is of the profit variety, the objective will be to maximize total payoff.

such a problem. Let us now turn our attention to delineating the relationship between the general linear programming problem and the special structure transportation problem.

9.2 THE TRANSPORTATION PROBLEM—A SPECIAL CASE

As we shall see in this section, the transportation problem has a specified structure that is a special case of the linear programming problem. We list again the three components of a general linear programming problem: (1) a linear objective function, (2) a set of linear structural constraints, and (3) a set of nonnegativity constraints. Let us construct a general linear programming problem that consists of m structural constraints and n real variables. Two equivalent representations of the general problem are as follows:

Maximize

$$F(X) = \sum_{j=1}^{n} c_j x_j$$

subject to the constraints

$$\sum_{j=1}^{n} a_{ij} x_j \leqslant b_i, \; i = 1, 2, \ldots, m$$

and

$$x_j \geqslant 0 \quad j = 1, 2, \ldots, n$$

Maximize

$$F(x) = c_1 x_1 + c_2 x_2 + \cdots + c_n x_n$$

subject to the constraints

$$a_{11} x_1 + a_{12} x_2 + \cdots + a_{1n} x_n \leqslant b_1$$
$$a_{21} x_1 + a_{22} x_2 + \cdots + a_{2n} x_n \leqslant b_2$$
$$\vdots \qquad \vdots \qquad \vdots$$
$$a_{m1} x_1 + a_{m2} x_2 + \cdots + a_{mn} x_n \leqslant b_m$$

and

$$x_1, \quad x_2, \ldots, \quad x_n \geqslant 0$$

where

c_j = set of profit or cost coefficients in the objective function
x_j = set of real variables that represent competing candidates or activities
a_{ij} = input–output coefficients that represent state of technology
b_i = set of right-hand-side constants, reflecting maximum resource capacities or minimum requirements

We have shown the structural constraints in the form of "less than or equal to" inequalities. They can, of course, be simple equations or inequalities of the "greater than or equal to" type. Thus, in addition to x_j (the real variables), we often utilize slack, surplus, and artificial variables to fit the problem to the simplex tableau format presented in Chapter 7.

Two remarks must be made at this time in reference to the general linear programming problem. First, the input–output coefficients a_{ij} are not restricted to any particular value or values. For example, a particular a_{ij} may be specified to have a value of 10, -20, 1, or 0. Second, no restrictions are imposed regarding the homogeneity of units among the various inequalities that represent the structural constraints. Of the given constraints, some may refer to available capacities of machines that perform different kinds of operations, while others may specify different types of, say, chemical characteristics. In other words, the units of any one constraint may not be the same as those of the other constraints and hence may not be interchangeable with the units of any other constraint. We can illustrate this by referring back to the vitamin problem of Chapter 7, in which the two structural constraints were concerned with different types of vitamins.

The transportation problem, in comparison with the general linear programming problem, restricts the values that can be assigned to the input–output coefficients and limits the constraints to only one type of unit, so that they are homogeneous and hence interchangeable. In particular, the general linear programming problem can be reduced to what is called a transportation problem if (1) the input–output coefficients a_{ij}'s (coefficients of the structural or real variables in the constraints) are restricted to the values 0 and 1, and (2) there exists a homogeneity of units among the constraints. Hence the transportation problem is a special case of the linear programming problem.

Let us now formulate a typical transportation problem that involves three origins and four destinations.

9.3   AN ILLUSTRATIVE EXAMPLE

A manufacturing concern has three plants located in three different cities, all producing the same product. The total supply potential of the firm is absorbed by four large customers. Let us identify the three plants as $O_1, O_2,$ and O_3, and the customers as $D_1, D_2, D_3,$ and D_4. The relevant data on plant capacities, destination requirements, and shipping costs for individual shipping routes are recorded, in general terms, in Table 9.1.*

Note that the first subscript in each symbol used in Table 9.1 refers to the specific origin and the second subscript to the particular destination. For example, c_{12} is the cost of shipping 1 unit of goods from origin O_1 to destination D_2, and the variable x_{34} is the quantity to be shipped from origin O_3 to destination D_4. Origin capacities and destination requirements are given along the outside (rims) of Table 9.1 and are usually referred to as *rim requirements*. Our problem is to choose the strategy (a particular program of shipping) that will satisfy the rim requirements at a minimum total cost.

* The matrix of the transportation problem in Table 9.1 has three rows and four columns and hence is not a square matrix. This is to emphasize the point that a transportation problem is not restricted to a square matrix. As we shall see in Chapter 10, the "assignment model" is restricted to a square matrix in the sense that one origin cannot simultaneously be associated with more than one destination. In the transportation problem, one origin can simultaneously supply goods to many destinations.

Table 9.1

Origin	Destination				Origin Capacity per Time Period
	D_1	D_2	D_3	D_4	
O_1	c_{11} x_{11}	c_{12} x_{12}	c_{13} x_{13}	c_{14} x_{14}	b_1
O_2	c_{21} x_{21}	c_{22} x_{22}	c_{23} x_{23}	c_{24} x_{24}	b_2
O_3	c_{31} x_{31}	c_{32} x_{32}	c_{33} x_{33}	c_{34} x_{34}	b_3
Destination Requirement per Time Period	d_1	d_2	d_3	d_4	

c_{ij} = cost of shipping 1 unit of goods from ith origin to jth destination. x_{ij} = number of units to be shipped from ith origin to jth destination. $\sum b_i = \sum d_j$. That is, total origin capacities equal total destination requirements.

9.3.1 Analysis of the Problem

The transportation problem described above, like the general linear programming problem, consists of three components. First, we can formulate a linear objective function that is to be minimized. This function will represent the total shipping cost of all the goods to be sent from the three origins to the four destinations. Second, we can write a set of linear structural constraints. There are seven structural constraints for this problem: three reflecting the row or origin capacities and four reflecting the column or destination requirements. Hence three constraints (one for each row) relate the origin capacities and the goods to be received by different destinations. These are called *row* or *capacity constraints*. The other four constraints (one for each column) will specify the relationships between destination requirements and the goods to be shipped from different origins. These are called *column* or *requirement constraints*. Third, we can specify a set of nonnegativity constraints for the variables x_{ij}. They will state that no negative shipment is permitted. The general correspondence between a typical linear programming problem and the transportation problem is thus complete.

The three component parts of our transportation problem are given next:

Minimize $\quad F(X) = c_{11}x_{11} + c_{12}x_{12}$

$$+ c_{13}x_{13} + c_{14}x_{14} + c_{21}x_{21} + c_{22}x_{22} + c_{23}x_{23}$$
$$+ c_{24}x_{24} + c_{31}x_{31} + c_{32}x_{32} + c_{33}x_{33} + c_{34}x_{34}$$

subject to the constraints

$$
\begin{aligned}
x_{11} + x_{12} + x_{13} + x_{14} &= b_1 \\
x_{21} + x_{22} + x_{23} + x_{24} &= b_2 \\
x_{31} + x_{32} + x_{33} + x_{34} &= b_3 \\
x_{11} \qquad + x_{21} \qquad + x_{31} \qquad &= d_1 \\
x_{12} \qquad + x_{22} \qquad + x_{32} \qquad &= d_2 \\
x_{13} \qquad + x_{23} \qquad + x_{33} \qquad &= d_3 \\
x_{14} \qquad + x_{24} \qquad + x_{34} &= d_4
\end{aligned}
$$

and

$$x_{ij} \geqslant 0 \qquad i = 1, 2, 3 \quad j = 1, 2, 3, 4$$

Based on the preceding explanation, we can write the general transportation model as follows:

Minimize $$\sum_{i=1}^{m} \sum_{j=1}^{n} c_{ij}x_{ij} \qquad (9.1)$$

subject to the constraints

$$\sum_{j=1}^{n} x_{ij} = b_i \qquad i = 1, 2, \ldots, m \qquad (9.2)$$

$$\sum_{i=1}^{m} x_{ij} = d_j \qquad j = 1, 2, \ldots, n \qquad (9.3)$$

and

$$x_{ij} \geqslant 0 \qquad b_i \geqslant 0 \qquad d_j \geqslant 0 \qquad (9.4)$$

and the additional requirement that the sum of origin capacities equals the sum of the destination requirements. That is,

$$\sum_{i=1}^{m} b_i = \sum_{j=1}^{n} d_j \qquad (9.5)$$

The transportation problem given by (9.1) to (9.5) is called the *balanced* transportation problem. If the sum of origin capacities is not equal to the sum of destination requirements, the problem is *unbalanced*.

The unbalanced problem can be converted to a balanced problem by the simple device of adding a *dummy* origin (to supply the excess requirement) or a dummy destination (to absorb the excess capacity). In the row (representing dummy origin) or the column (representing dummy destination), all transportation costs are assumed to be zero. We illustrate the conversion of the unbalanced problem to a balanced problem in Section 9.12.

It is obvious from the expanded form of our transportation problem of Table 9.1 that all its input–output coefficients are either 1 or zero. Furthermore, the problem can certainly be fed into the simplex format and solved by the simplex algorithm. But this would be a rather lengthy process and would not add anything new to our knowledge of the simplex method. As can be verified, the initial simplex tableau for a simple 3×4 transportation problem would consist of 7 rows and 19 columns (12 real variables and 7 artificial variables). Fortunately, however, a simple, routine, and more efficient method of solving such problems has been developed. It is fittingly called the transportation method. Whenever a given linear programming problem can be placed in the transportation framework, it is far simpler to solve it by the transportation method than by the simplex method. Before we describe and develop the transportation method, let us comment on certain characteristics of the transportation problem and its solution.

First, a little reflection will show that for the transportation problem of Table 9.1, only six rather than seven structural constraints need be specified. In view of the fact that the sum of the origin capacities equals the sum of the destination requirements ($\sum b_i = \sum d_j$), any solution satisfying six of the seven constraints will automatically satisfy the remaining constraint. In general, therefore, if m represents the number of rows and n represents the number of columns in a given transportation problem, we can state the problem completely with $m + n - 1$ equations. This means that a basic feasible solution of a transportation problem has exactly $m + n - 1$ positive components (as compared to $m + n$ positive components required for a basic feasible solution for the general linear programming problem having $m + n$ structural constraints).

Second, if origin capacities equal destination requirements, it is always possible to design an initial basic feasible solution in such a manner that the rim requirements are satisfied. This can be accomplished in one of the ways described in Section 9.5.

9.4 APPROACH OF THE TRANSPORTATION METHOD

The transportation method consists of three basic steps. The first step involves making the initial shipping assignment in such a manner that a basic feasible solution is obtained. This means that $m + n - 1$ cells (routes) of the transportation matrix are used for shipping purposes. The cells having the shipping assignment will be called *occupied cells*, while the remaining cells of the transportation matrix will be referred to as *empty cells*.

The purpose of the second step is to test the optimality of the solution. This is done by determining the opportunity costs associated with the empty cells. The opportunity costs of the empty cells can be calculated individually for each cell or simultaneously for the whole matrix. If the opportunity costs of all the empty cells are nonpositive, we can be confident that an optimal solution has been obtained.* On the other hand, if even a single empty cell has a positive opportunity cost, we proceed to step 3.

The third step involves determining a new and better basic feasible solution. This involves using an heretofore empty cell so that the *highest rate of improvement* in the objective function is obtained. Once this new basic feasible solution has been obtained, we repeat steps 2 and 3 until an optimal solution has been designed.

The remaining sections of this chapter are devoted to illustrating the development and application of the above-mentioned approach to the solution of a given transportation problem.

9.5 METHODS OF MAKING THE INITIAL ASSIGNMENT

The first step in the transportation method, as stated above, consists in making an initial assignment in such a manner that a basic feasible solution (number of occupied cells equals $m + n - 1$) is obtained. Of the various methods of making such an assignment, these three are discussed most often: (1) northwest-corner rule, (2) least-cost method, and (3) Vogel's approximation method (or the penalty method). We shall illustrate them with respect to the transportation problem shown in Table 9.2.

9.5.1 Northwest-Corner Rule

According to this rule, the first allocation is made to the cell occupying the upper left-hand (northwest) corner of the matrix. Further, this allocation is of such a magnitude that either the origin capacity of the first row is exhausted, or the destination requirement of the first column is satisfied, or both. If the origin capacity of row 1 is exhausted first, we move down the first column and make another allocation which either exhausts the origin capacity of row 2 or satisfies the remaining destination requirement of column 1. On the other hand, if the first allocation completely satisfies the destination requirement of column 1, we move to the right in row 1 and make a second allocation which either exhausts the remaining capacity of row 1 or satisfies the destination requirement of column 2, and so on. In this manner,

* The transportation problem falls under the category of decision making under certainty (models that deal with decision making under certainty are referred to as *deterministic models*; see Section 2.3). Hence the optimal solution must not contain any positive opportunity costs (opportunity cost was defined in Section 7.3.2).

Table 9.2

Origin	Destination				Origin Capacity per Time Period
	D_1	D_2	D_3	D_4	
O_1	12	4	9	5	55
O_2	8	1	6	6	45
O_3	1	12	4	7	30
Destination Requirement per Time Period	40	20	50	20	130

starting from the upper left-hand corner (i.e., northwest corner) of the given transportation matrix, satisfying the individual destination requirements, and exhausting the origin capacities one at a time, we move toward the lower right-hand corner until all the rim requirements are satisfied. It should be noted that when we follow the northwest-corner rule, we pay no attention to the relative costs of the different routes while making the first assignment.

For the transportation problem of Table 9.2, application of the north-west-corner rule dictates that we first "load" or "fill" cell O_1D_1, which lies in the upper left-hand (northwest) corner. The product requirement of D_1 is 40 units, and the capacity of O_1 is 55 units; the lower of these two numbers (i.e., 40) is placed in cell O_1D_1. This means that the requirement of D_1 is fully satisfied, but we still have 15 units (55–40) of unused capacity at O_1. Thus we move to the right of cell O_1D_1 in the first row. At this stage we note that the destination requirement of column D_2 is 20 units. Knowing that 15 units of capacity O_1 are still unused, we route all 15 units to destination D_2 (place 15 in cell O_1D_2). This completely exhausts the capacity O_1, but column D_2 still needs 5 units (20 − 15) to satisfy its requirement. Thus we move down column D_2 and supply these 5 units from capacity O_2 (place 5 in cell O_2D_2). This leaves 40 units of unused capacity at O_2; these are routed to D_3 (place 40 in cell O_2D_3). The remaining requirement of 10 units (50 − 40), for D_3 is supplied from O_3 (place 10 in cell O_3D_3). This leaves 20 units of unused capacity at O_3; these are routed to D_4 (place 20 in cell O_3D_4). The entire table has now been loaded, resulting in the initial program given in Table 9.3. The circled numbers in the table give the number of units shipped from a

Table 9.3. Initial Assignment by Northwest-Corner Rule

Origin	Destination				Total
	D_1	D_2	D_3	D_4	
O_1	12 ⑩40	4 ⑮15	9	5	55
O_2	8	1 ⑤5	6 ㊵40	6	45
O_3	1	12	4 ⑩10	7 ⑳20	30
Total	40	20	50	20	130

particular origin to a certain destination. The cells in which these circled numbers are entered are the occupied cells. The rest of the cells are the empty cells. The occupied cells correspond to the basic variables; the empty cells correspond to the nonbasic variables of the simplex method.

It is to be observed that the number of occupied cells is

$m + n - 1 = 3 + 4 - 1 = 6$

$$= \text{(number of rows + number of columns} - 1)$$

The solution at this stage is, therefore, not degenerate.*

The total cost of this assignment is

$$40 \times 12 + 15 \times 4 + 5 \times 1 + 40 \times 6 + 10 \times 4 + 20 \times 7 = \$965$$

It should be noted that the last allocation (cell O_3D_4) simultaneously satisfied the requirement of column D_4 and exhausted the capacity of O_3. This is the normal situation in the last allocation made by the northwest-corner rule; and if this occurs only in the last allocation, we can be certain of having a basic feasible solution. However, if any allocation previous to the last allocation happens to be such that it simultaneously satisfies the requirement of some destination and exhausts the capacity of some origin, then the number of occupied cells will be less than $m + n - 1$. This will mean that we have a degenerate basic feasible solution.† The reader should try to make an

---

* A basic feasible solution for a transportation problem requires exactly $m + n - 1$ positive components. Thus, whenever a transportation program has $m + n - 1$ occupied cells, the solution is not degenerate.

† A method for resolving degeneracy in transportation problems is discussed in Section 9.14.

initial assignment in Table 9.2 by following the northwest-corner rule after having changed the destination requirements of D_2 and D_3 to 15 and 55 units, respectively.

9.5.2 The Least-Cost Method

The *least-cost method* starts by making the first allocation to that cell whose shipping cost per unit is lowest. This lowest-cost cell is loaded or filled as much as possible in view of the origin capacity of its row and the destination requirement of its column. Then we move to the next lowest-cost cell and make an allocation in view of the remaining capacity and requirement of its row and column, and so on. Should there be a tie for lowest-cost cell during any allocation, we can exercise judgement in breaking the tie or we can arbitrarily choose a cell for allocation.

Let us illustrate the least-cost method for the transportation problem of Table 9.2. We note that cells $O_2 D_2$ and $O_3 D_1$ each have a shipping cost of $1 per unit. Thus there is a tie for the first allocation. We arbitrarily choose cell $O_3 D_1$ for the first allocation and route 30 units from O_3 to D_1. This means that the capacity of O_3 is fully utilized (cross off row O_3 with a light pencil). The second allocation is made to cell $O_2 D_2$, and we ship 20 units through this route (place 20 in cell $O_2 D_2$ and cross off column D_2). Of those remaining, cell $O_1 D_4$ has the lowest cost, and we route 20 units through $O_1 D_4$ (place 20 in cell $O_1 D_4$ and cross off column D_4). Of those remaining, cell $O_2 D_3$ has the lowest cost, and we route 25 units (the remaining capacity of row O_2) through cell $O_2 D_3$ (place 25 in cell $O_2 D_3$ and cross off row O_2). We are now left with 35 units at O_1, while D_1 and D_3 still require 10 and 25 units, respectively. Hence we route 10 units through $O_1 D_1$ and 25 units through $O_1 D_3$. All the rim requirements have now been satisfied, and we have the initial assignment in Table 9.4. The dotted lines crossing the cost squares (c_{ij}'s) have been numbered to show the order in which different rows and columns were crossed off, as an aid to making the initial assignment by the least-cost method.

It is to be observed that the number of occupied cells is 6 (i.e., $m + n - 1$), and thus we have a basic feasible solution. The total cost of this assignment is $645. It is $320 less than the cost of the initial solution obtained by the northwest-corner rule.

9.5.3 Vogel's Approximation Method (or the Penalty Method)

According to *Vogel's approximation method* (or *the penalty method*), a difference column and a difference row representing the difference between the costs of the *two cheapest* routes for each origin and destination are computed. Each individual difference can be thought of as a penalty for not using the cheapest route. After all such penalty ratings have been computed for the given data, the highest difference or penalty rating is identified. Then

Table 9.4. Initial Assignment by the Least-Cost Method

Origin	Destination				Total
	D_1	D_2	D_3	D_4	
O_1	12 (10)	4	9 (25)	5 (20)	5̸5̸ 3̸5̸ 1̸0̸ 0
O_2	8	1 (20)	6 (25)	6	4̸5̸ 2̸5̸ 0 ·····4th
O_3	1 (30)	12	4	7	3̸0̸ 0 ·····1st
Total	4̸0̸ 1̸0̸ 0	2̸0̸ 0	5̸0̸ 2̸5̸ 0	2̸0̸ 0	
	6th	2nd	5th	3rd	

the lowest-cost cell in that row or column in which the highest penalty rating was placed is the cell to which the first assignment is made.* This assignment either exhausts the capacity of an origin, or meets the requirement of a destination, or both. The particular row or column that has been thus satisfied is removed from the transportation matrix. The process is then repeated until an initial assignment has been obtained. The penalty method has the disadvantage of necessitating some computational work before the initial program is obtained, but it usually results in the attainment of the optimal program in fewer iterations than are required when the initial program is obtained by using the northwest-corner rule.

The mechanics for obtaining the initial assignment for the transportation problem of Table 9.2 by the penalty method is illustrated in Tables 9.5a through 9.5e. In Table 9.5a, the highest difference or penalty rating is 7, and this falls under column D_1. The first allocation, therefore, must be made to that cell in column D_1 which has the lowest shipping cost. Since cell $O_3 D_1$ has the lowest shipping cost in that column, we now compare the capacity of O_3 (30 units) with the requirement of D_1 (40 units). The lower of the two numbers (i.e., 30) is placed in cell $O_3 D_1$. This means that the capacity of O_3 has been fully utilized, and row O_3 can be removed temporarily from the transportation matrix. Column D_1, however, cannot be removed, since we still need 10 units to satisfy its requirements fully.

---

* Should there be a tie for highest penalty rating or difference value, we can arbitrarily choose one to break the tie.

Table 9.5a

	D_1	D_2	D_3	D_4	Capacity	Difference or Penalty Column
O_1	12	4	9	5	55	1
O_2	8	1	6	6	45	5
O_3	1 ⑳(30)	12	4	7	~~30~~ 0	3
Requirement	~~40~~ 10	20	50	20		
	D_1	D_2	D_3	D_4		
O_1	12	4	9	5		
O_2	8	1	6	6		
O_3	1	12	4	7		
Difference or Penalty Row	7 ✓	3	2	1		

Table 9.5b

	D_1	D_2	D_3	D_4		Difference or Penalty Column		D_1	D_2	D_3	D_4	Capacity
O_1	12	4	9	5		1	O_1	12	4	9	5	55
O_2	8	1	6	6		5 √	O_2	8	1 (20)	6	6	4̶5̶ 25
Difference or Penalty Row	4	3	3	1			Require-ment	10	2̶0̶ 0	50	20	

Table 9.5c

	D_1	D_3	D_4	Difference or Penalty Column
O_1	12	9	5	4 ✓
O_2	8	6	6	0
Difference or Penalty Row	4	3	1	

	D_1	D_3	D_4	Capacity
O_1	12	9	5 (20)	5̶5̶ 35
O_2	8	6	6	25
Requirement	10	50	2̶0̶ 0	

Table 9.5d

	D_1	D_3	Difference or Penalty Column		D_1	D_3	Capacity
O_1	12	9	3	O_1	12	9	35
O_2	8	6	2	O_2	8 ⑩	6	~~25~~ 15
Difference or Penalty Row	4 √	3		Requirement	~~10~~ 0	50	

Table 9.5e

	D_3	Capacity
O_1	9 (35)	~~35~~ 0
O_2	6 (15)	~~15~~ 0
Requirement	~~50~~ 0	

We now have arrived at Table 9.5b. The process of computing penalties is repeated, and in Table 9.5b we note that the highest penalty falls in row O_2. Therefore, we make an assignment in the lowest-cost cell of row O_2. This assignment (place 20 in cell $O_2 D_2$) is such that column D_2 can now be removed from the matrix, and we proceed to Table 9.5c. By repeating the same process in Tables 9.5c to 9.5e, we finally obtain the assignment in Table 9.6. Observe that the number of occupied cells is $m + n - 1 = 3 + 4 - 1 = 6$. The initial solution is therefore a basic feasible solution, and the problem at this stage is not degenerate. The total cost of this assignment is $635, considerably less than the total cost associated with the initial program obtained via the northwest-corner rule.

Table 9.6. Initial Assignment by the Penalty Method

Origin	Destination				Total
	D_1	D_2	D_3	D_4	
O_1	12	6	9 (35)	5 (20)	55
O_2	8 (10)	1 (20)	6 (15)	6	45
O_3	1 (30)	12	4	7	30
Total	40	20	50	20	130

9.6 TEST FOR OPTIMALITY

Once the initial basic feasible solution has been obtained, our next step is to test its optimality. There are two methods for testing optimality in transportation problems. In one method, test of optimality is conducted by directly calculating the opportunity cost of each empty cell. This is the *stepping-stone method*. In the stepping-stone method, each empty cell (associated with a nonbasic variable) is evaluated by transferring into it one unit from an occupied cell (associated with a basic variable). The effect on the objective function of such an exchange is then measured in terms of opportunity cost (i.e., $C_j - Z_j$). Similarly, opportunity costs for all empty cells are calculated, one at a time. Even if a single opportunity cost (or *cell evaluator*) is positive, our program is not optimal. A new and better program is then designed. We illustrate the stepping-stone method in Section 9.8.

The second method of testing optimality is called the *modified distribution method (MODI)*. This method is based on the concept of dual variables, which are used for evaluating the empty cells of a given program. As we shall see in Section 9.9, there is a simple way of assigning values to the dual variables associated with a given solution. Once this is done, the opportunity costs of all the empty cells of a program can easily be calculated. As compared to the stepping-stone method, the MODI method of testing optimality is simpler and more efficient.

In both the stepping-stone method and the MODI method, after the nonoptimality of a program has been established, the *most favorable* empty cell is identified. Then a new program is designed so that this most favorable empty cell becomes one of the occupied cells.

9.7 DESIGN A NEW AND BETTER PROGRAM

The procedure for designing the new and better program (step 3 of Section 9.4), as we shall see in Sections 9.8 and 9.9, is based on the overall constraint that *all* the rim requirements are fully met. As in the simplex method, all we do is exchange a nonbasic variable (most favorable empty cell) for at least one basic variable (one of the occupied cell). While making the transfer of goods to accomplish this, we must make certain that row as well as column requirements remain satisfied.

We now present the stepping-stone method (Section 9.8) and the MODI method (Section 9.9) of testing optimality. In each method we shall continue to design new and better programs until an optimal program has been obtained. In other words, we shall execute steps 2 and 3 of the transportation method (as described in Section 9.4) in Section 9.8 as well as in Section 9.9.

9.8 THE STEPPING-STONE METHOD

To illustrate the stepping-stone method, we shall first solve the very simple transportation problem given in Table 9.7. The method will then be used in deriving the optimum solution to our problem of Table 9.2. The purpose of solving the simple problem of Table 9.7 is to familiarize the reader with the terminology and rationale of the stepping-stone method.

Table 9.7

Origin	Destination		Origin Capacity per Time Period
	D_1	D_2	
O_1	2	2	1,000
O_2	1	2	600
Destination Requirement per Time Period	900	700	1,600 / 1,600

Following the northwest-corner rule, we obtain the initial program in Table 9.8. The circled numbers within the body of the matrix refer to the specific allocations of the first program. This program calls for shipping 900 units from O_1 to D_1, 100 units from O_1 to D_2, and 600 units from O_2 to D_2. Obviously, this program satisfies all the rim requirements. Note further that the number of occupied cells is 3, which is 1 less than the sum of the numbers of rows and columns. In other words, the number of occupied cells in this program equals $m + n - 1$.* Thus we have a basic feasible solution.

---

* As we have established previously, only $m + n - 1$ equations are needed to state a transportation problem. The present problem can be stated with the following equations:

$$x_{11} + x_{12} = 1,000$$
$$x_{21} + x_{22} = 600$$
$$x_{11} + x_{21} = 900$$

Naturally, letting $x_{21} = 0$ gives a solution in which $x_{11} = 900$, $x_{22} = 600$, and $x_{12} = 100$.

Table 9.8

Origin	Destination		Total
	D_1	D_2	
O_1	2 (900)	2 (100)	1,000
O_2	1	2 (600)	600
Total	900	700	

9.8.1 Test for Optimality

Is the preceding program an optimal program? To answer this question, we must apply step 2, that is, determine the opportunity costs of the empty cells. Insofar as the transportation model involves decision making under certainty, we know that an optimal solution must not incur any positive opportunity cost. Thus, to determine whether any positive opportunity cost is associated with a given program, we must test the empty cells (cells representing routes not used in the given program) of the transportation matrix for the presence or absence of opportunity cost. The absence of positive opportunity cost in all empty cells will indicate that an optimal solution has been obtained. If, on the other hand, even a single empty cell has a positive opportunity cost, the given program is not the optimal program and hence should be revised.*

Let us examine our first program in Table 9.8 in view of the preceding discussion. Since cell O_2D_1 in this program is empty, we wish to determine whether there is an opportunity cost associated with it. This is accomplished by shifting 1 unit of goods to cell O_2D_1, making other shifts necessary to satisfy the rim requirements, and then finding the cost consequence of these changes. Let us shift 1 unit from cell O_2D_2 to cell O_2D_1. This shift will necessitate the changes noted in Table 9.9 to keep the rim requirements satisfied. These changes are associated with the following cost consequence or cost change:

$$-2 + 1 - 2 + 2 = -1 \text{ dollar}$$

Since the shifting of 1 unit to O_2D_1 yields a negative cost change, it is obviously a desirable shift. The fact that the transfer of 1 unit to cell O_2D_1 resulted in a net cost change of -1 dollar indicates that the opportunity cost

* It will be recalled that the test for optimality in the simplex method was also based on the concept of opportunity cost.

Table 9.9

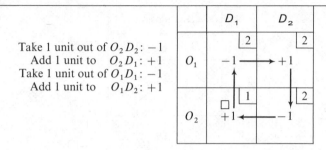

Take 1 unit out of $O_2 D_2$: -1
Add 1 unit to $O_2 D_1$: $+1$
Take 1 unit out of $O_1 D_1$: -1
Add 1 unit to $O_1 D_2$: $+1$

of *not* including cell $O_2 D_1$ in the first program is $+1$ dollar per unit of shipment. The empty cell $O_2 D_1$ must, therefore, be included in a new and improved program.

9.8.2 Design a New and Better Program

Having discovered that the opportunity cost of the empty cell $O_2 D_1$ is positive, we must next obtain a new basic feasible solution. This is done by designing a new improved program in which cell $O_2 D_1$ is included in the shipping strategy. Let us make the improvement by shifting just 1 unit from cell $O_2 D_2$ to cell $O_2 D_1$. The revised program is given in Table 9.10. The shift of 1 unit from cell $O_2 D_2$ to cell $O_2 D_1$ means that we are left with 599 units in cell $O_2 D_2$. This change, it should be noted, has not violated the capacity constraints of either row 1 or row 2. But what about the requirement constraints of columns 1 and 2? With the above change, we now have 1 unit in $O_2 D_1$, 900 units in $O_1 D_1$, 599 units in $O_2 D_2$, and 100 units in $O_1 D_2$. In other words, column 1 has 901 units (one more unit than the requirement of column 1), and column 2 has 699 units (one less unit than the requirement of column 2). Clearly, this situation can be remedied by shifting 1 unit from cell $O_1 D_1$ to cell $O_1 D_2$, a change that will simultaneously satisfy the row and column requirements.

Table 9.10

| | First Program | | Revised Program (with 1 Unit Shifted to $O_2 D_1$) | |

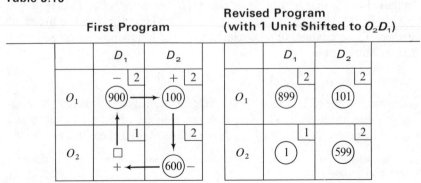

The revised program, with 1 unit shifted to cell $O_2 D_1$, is shown in Table 9.10. The change in the program effected by the introduction of 1 unit to cell $O_2 D_1$, as we established earlier, reduces the total shipping cost by \$1. Insofar as we gain this advantage each time a unit is shifted to cell $O_2 D_1$, we must shift to cell $O_2 D_1$ as many units as possible. As the closed loop (plus and minus signs connected by arrows) of Table 9.10 shows, we cannot shift more than 600 units to $O_2 D_1$, for the allocation of more than 600 units to cell $O_2 D_1$ would certainly violate the capacity constraint of row O_2.

Our second program (a better basic feasible solution), the result of the above discussion, appears in Table 9.11. Is this the optimal allocation? The answer to this question can be obtained by testing the opportunity cost of cell $O_2 D_2$, which is now the only empty cell. The answer is in the affirmative, since the opportunity cost of cell $O_2 D_2$ is not positive. This can be verified by shifting 1 unit to cell $O_2 D_2$ and noting that the net cost consequence of such a shift is $+1$ dollar $(+2 - 2 + 2 - 1)$. The opportunity cost, being the negative of the corresponding net cost change, is therefore negative. Hence the assignment of Table 9.11 gives an optimal solution with a total shipping cost of \$2,600. No other program for this problem can result in a lower total shipping cost.

Table 9.11. Second and Optimal Program

	D_1	D_2	Total
O_1	2 (300)	2 (700)	1,000
O_2	1 (600)	2	600
Total	900	700	

Let us recapitulate briefly the method of attack followed in solving this transportation problem. First, we designed a basic feasible solution by following the northwest-corner rule for making an initial assignment.*

---

* In most cases, the initial assignment made by following the northwest-corner rule will be such that the number of occupied or filled cells equals $m + n - 1$, where m is the number of rows and n is the number of columns. When this happens, we have a basic feasible solution. If the number of occupied cells in the initial assignment is less than $m + n - 1$, the problem is said to be degenerate at the very beginning. This type of degeneracy, as well as the degeneracy occurring during the solution stages, can easily be resolved by judicious placement of *a small-number*, epsilon, in the empty cell(s). The epsilon is to be placed in that cell(s) which will help complete the different loops for all the empty cells. See Section 9.14 and Problem 9.8 (initial assignment by the least cost method).

Second, having obtained a basic feasible solution, we proceeded to determine the opportunity cost of the empty cell in order to determine whether the first program was an optimal program. The method employed to determine the opportunity cost of the empty cell consisted in (1) drawing a closed loop that passed through the empty cell and the adjacent occupied cells with proper plus and minus signs at the corners of the loop,* (2) shifting 1 unit of goods to the empty cell (accomplished by the addition of 1 unit to all those cells in which fell a plus sign of the closed loop and by the subtraction of 1 unit from all cells in which fell a minus sign), (3) determining the net cost change associated with shifting 1 unit to the empty cell, and (4) taking the negative of the net cost change in (3) to find the opportunity cost of the empty cell. In the simple transportation problem of Table 9.7, we had to find the opportunity cost of only one empty cell. In a problem of larger dimensions, the opportunity costs of *all* the empty cells must be determined by this procedure. The point to emphasize here is that a separate closed loop must be established for every empty cell (in the stepping-stone method) *before* the opportunity costs of the empty cells can be determined.

Third, after ascertaining that the opportunity cost of the empty cell was positive, we changed the initial program by filling the empty cell $(O_2 D_1)$ as much as possible in view of the rim requirements.† The revision of the given program was guided by the plus and minus signs of the closed loop. The smallest of the numbers in the cells in which minus signs of the closed loop appeared (600) gave the total number of units to be shifted to the empty cell. Ths shifting was accomplished by adding this number (600) to all the cells that contain the plus signs of the loop and subtracting it from all the cells that contain the minus signs of the loop. These changes gave us our new basic feasible solution.

Finally, we tested the only empty cell $(O_2 D_2)$ of the second program and found that its opportunity cost was not positive. Therefore, we came to the conclusion that an optimal solution to our problem had been obtained.

---

* While tracing this closed loop, one should start with the empty cell being evaluated and draw an arrow from that empty cell to an occupied cell in the same row or column. Then a plus sign is placed in the empty cell and a negative sign in the occupied cell to which the arrow was drawn. Next, one moves horizontally or vertically (never diagonally) to another occupied cell, and so on, until one is back to the original empty cell. At each turn of the loop, plus and minus signs are placed alternately. Further, there is the important restriction that there are exactly one positive terminal and exactly one negative terminal in any row or column through which the loop happens to pass. Obviously, this restriction is imposed to ensure that the rim requirement will not be violated when the units are shifted (to obtain a new program) along this closed loop. Mechanically, this implies that during the tracing of the closed loop, right-angle turns must be made only at the occupied cells. The starting point of a closed loop is identified in this book by the symbol □.

† The fact that the opportunity cost of even a single cell is positive indicates that an optimal solution has not been obtained and that the given program must be revised to obtain a better basic feasible solution. Normally, the improved program will include that empty cell whose opportunity cost is highest. Since in this problem we had only one empty cell, $O_2 D_1$, the new program included that cell.

The procedure described above forms the core of the stepping-stone method. Although the transportation problem that we solved was represented only by a 2×2 matrix (Table 9.7), the stepping-stone method may be applied to any $m \times n$ matrix.

Suppose that our initial assignment in a 4×5 (4 rows and 5 columns) transportation problem results in 8 occupied cells and 12 empty cells. To test the optimality of this program, and then to revise it, we must calculate the opportunity cost of each of the 12 empty cells. If we discover that the initial program can be improved, we revise the program by including that empty cell whose opportunity cost is highest. Note that regardless of the number of empty cells that have positive opportunity costs, only one cell at a time is included in the new program.

We shall now apply the stepping-stone method to the problem of Table 9.2.

STEP 1. Obtain an initial basic feasible solution.

An initial basic feasible solution for a given transportation problem may be obtained by following the northwest-corner rule, by the least-cost method, or by the penalty method. It will be recalled that for the transportation problem of Table 9.2 we obtained three different basic feasible solutions, given in Tables 9.3, 9.4, and 9.6. Of these, let us take the basic feasible solution of Table 9.4 (initial assignment by the least-cost method) as the starting point for obtaining the optimal solution by the stepping-stone method. The data of Table 9.4 are repeated in Table 9.12.

Table 9.12. First Program

Origin	Destination				Total
	D_1	D_2	D_3	D_4	
O_1	12 ⑩	4	9 ㉕	5 ⑳	55
O_2	8	1 ⑳	6 ㉕	6	45
O_3	1 ㉚	12	4	7	30
Total	40	20	50	20	

Table 9.13

Empty Cell	Closed Loop	Net Cost Change	Opportunity Cost	Action
O_1D_2	$+O_1D_2 - O_1D_3 + O_2D_3 - O_2D_2$	$+4 - 9 + 6 - 1 = 0$	0	Indifferent
O_2D_1	$+O_2D_1 - O_1D_1 + O_1D_3 - O_2D_3$	$+8 - 12 + 9 - 6 = -1$	$+1$	Candidate for inclusion in next program
O_2D_4	$+O_2D_4 - O_1D_4 + O_1D_3 - O_2D_3$	$+6 - 5 + 9 - 6 = +4$	-4	Do not include in next program
O_3D_2	$+O_3D_2 - O_3D_1 + O_1D_1 - O_1D_3 + O_2D_3 - O_2D_2$	$+12 - 1 + 12 - 9 + 6 - 1 = +19$	-19	Do not include in next program
O_3D_3	$+O_3D_3 - O_1D_3 + O_1D_1 - O_3D_1$	$+4 - 9 + 12 - 1 = +6$	-6	Do not include in next program
O_3D_4	$+O_3D_4 - O_1D_4 + O_1D_1 - O_3D_1$	$+7 - 5 + 12 - 1 = +13$	-13	Do not include in next program

Check on Step 1: Since the number of occupied cells in this program equals $m + n - 1$, that is, $3 + 4 - 1 = 6$, this is indeed a basic feasible solution, and not a degenerate solution.

STEP 2. Test for optimality (determine the opportunity costs of the empty cells).

We repeat: In the stepping-stone method, a separate closed loop with proper plus and minus signs must be completed for each of the empty cells before the respective opportunity costs can be calculated.* Since our first program has a total of 6 empty cells, 6 different closed loops must be drawn. The opportunity cost associated with each empty cell is calculated in Table 9.13. An examination of these opportunity costs shows that cell $O_2 D_1$ is the only cell with a positive opportunity cost. Hence cell $O_2 D_1$ must be included in our next program.

A comment about the opportunity cost of zero (for empty cell $O_1 D_2$) is in order. An opportunity cost of zero associated with a particular empty cell at any stage of the problem solution indicates that if this cell is included in the next program, the total cost of the new program will be the same as that of the current program. Thus we list "Indifferent"† in the "Action" column of Table 9.13.

STEP 3. Design a new and better program.

Having ascertained that a given program is not an optimal program (one or more empty cells have positive opportunity cost), we next revise the given program to obtain a new basic feasible solution. The revised program must include that empty cell of the current program whose opportunity cost is highest. No choice is available here, since cell $O_2 D_1$ is the only empty cell that has a positive opportunity cost.

The revision of the first program is guided by the closed loop of the empty cell to be included (in this case cell $O_2 D_1$) and is as shown in Table 9.14. Since 10 is the smallest number in a negative cell in the closed loop, it is added to the cells that contain plus signs and subtracted from the cells that contain minus signs.

The next question is: Does our revised program represent an optimal solution? To answer this question, we have to repeat step 2, as discussed previously (i.e., calculate the opportunity cost of each empty cell). Should the result of step 2 indicate a nonoptimal solution, we would repeat step 3 and obtain another basic feasible solution.

Determination of the opportunity costs of all the empty cells of the second program (Table 9.14) will reveal that an optimal solution has indeed been derived. The reader is encouraged to verify this by the application of step 2 to Table 9.14.

---

* The reader should firmly grasp both the logic and the technique utilized in drawing these closed loops. Note the difference between the closed loop for cell $O_1 D_2$ and that for $O_3 D_2$.

† This is equivalent to the presence of alternative solutions with the same total cost.

232

Table 9.14

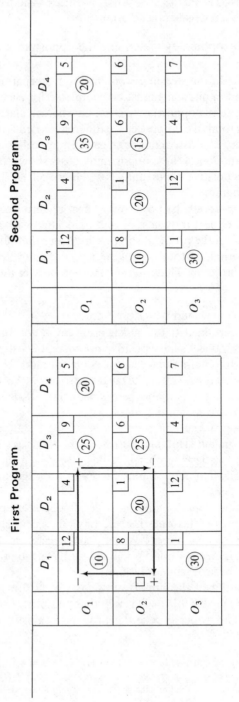

First Program

	D_1	D_2	D_3	D_4
O_1	12	4 ⑩→	9 ㉕ +	5 ⑳
O_2	8 ⑩	1 ⑳→	6 ㉕ –	6
O_3	1 ㉚	12	4	7

Second Program

	D_1	D_2	D_3	D_4
O_1	12	4	9 ㉟	5 ⑳
O_2	8 ⑩	1 ⑳	6 ⑮	6
O_3	1 ㉚	12	4	7

9.9 THE MODIFIED DISTRIBUTION METHOD

There are two major differences between the stepping-stone method and the
modified distribution method (MODI). The first concerns the manner in
which the test of optimality is conducted (i.e., how the opportunity costs of
the empty cells are calculated). In the stepping-stone method, we first draw
closed loops for each of the empty cells, and then use these loops to calculate
the opportunity costs (see Table 9.13). Then, having identified the most
favorable empty cell (i.e., cell with the highest opportunity cost), we use its
closed loop to effect the transfer of goods and obtain a new basic feasible
solution.

In the modified distribution method, as we shall illustrate below, test
of optimality is conducted (i.e., opportunity costs of the empty cells are
calculated) by the utilization of dual variables, *without first having to draw
closed loops for each empty cell*. As a matter of fact, the MODI method
easily permits us to identify the most favorable empty cell as soon as a set of
values for the dual variables has been chosen. Then a closed loop through the
most favorable empty cell is drawn so that a new basic feasible solution can
be obtained under the guidance of the closed loop. Note that in the modified
distribution method, we need draw only one closed loop *after* the highest-
opportunity-cost cell has been identified. Thus the procedure for calculating
the opportunity costs of the empty cells in MODI is independent of the
tracing of the loops.

We shall illustrate the mechanics and rationale of the modified-distribu-
tion method by solving the simple transportation problem of Table 9.7,
the data of which are reproduced in Table 9.15. The initial assignment made
by following the northwest-corner rule is given in Table 9.16.

Table 9.15

Origin	Destination		Origin Capacity per Time Period
	D_1	D_2	
O_1	2	2	1,000
O_2	1	2	600
Destination Requirement per Time Period	900	700	

Table 9.16

	D_1	D_2
O_1	$\boxed{2}$ $\textcircled{900}$	$\boxed{2}$ $\textcircled{100}$
O_2	$\boxed{1}$	$\boxed{2}$ $\textcircled{600}$

9.9.1 Test for Optimality (Determining the Opportunity Costs of the Empty Cells)

Having solved this problem by the stepping-stone method, we are aware of the fact that the transfer of 1 unit to cell $O_2 D_1$ (a closed loop for the empty cell was established in Table 9.9 before this transfer was made) results in a net cost change of -1 dollar. This, of course, means that the opportunity cost of not including cell $O_2 D_1$ in the first program is $+1$ dollar per unit of goods. Another method of reaching the same conclusion is via the determination of what may be called the *implied cost of an empty cell*, which sets an upper limit (in view of the existing program*) beyond which the inclusion of this cell in a new program is not an advantageous proposition. Let us explain this in connection with the transportation problem in Table 9.15.

In this case, cell $O_2 D_1$ is the only empty cell. We already know that one way to find its implied cost is by drawing a closed loop and determining the net cost consequence of shifting 1 unit of goods into $O_2 D_1$. Ignoring, for the time being, the actual shipping cost per unit via route $O_2 D_1$, we may calculate the net cost consequence of shifting 1 unit of goods into $O_2 D_1$ as

$$O_2 D_1 - O_1 D_1 + O_1 D_2 - O_2 D_2 = C_{21} - 2 + 2 - 2 = C_{21} - 2$$

Whatever the actual shipping cost per unit via cell $O_2 D_1$, it is obvious that the above shift is desirable only if the net cost change $(C_{21} - 2)$ is negative. It will be negative as long as the actual cost of $O_2 D_1$ is less than 2. The calculated upper limit for the actual cost of cell $O_2 D_1$ (in the existing program), beyond which the inclusion of this cell is not an advantageous proposition, is therefore 2. In other words, if the actual shipping cost via cell $O_2 D_1$ is greater than \$2 per unit, the shift is not desirable. On the other hand, if the actual shipping cost is less than \$2 per unit, the shift is desirable and cell $O_2 D_1$ should be included in the next program. The implied cost of the empty cell $O_2 D_1$ therefore is \$2 per unit.

---

* The implied cost of a given empty cell can change from one program to another, since the implied cost is indicative of the relative advantage or disadvantage of not using a given cell in a particular program.

Also, as we noted earlier, the negative of the net cost change involved in shifting 1 unit of goods to an empty cell gives the opportunity cost associated with the empty cell. For cell $O_2 D_1$,

$$\text{opportunity cost} = -(\text{net cost change}) = -(C_{21} - 2) = 2 - C_{21}$$

where C_{21} is the actual cost of shipment per unit via cell $O_2 D_1$. But, as we have just calculated, the implied cost of not using cell $O_2 D_1$ is \$2 per unit. Hence

$$\text{opportunity cost} = \text{implied cost} - \text{actual cost}$$

Substituting the actual shipping cost via cell $O_2 D_1$ (\$1) and the calculated implied cost of cell $O_2 D_1$ (\$2) in the preceding expression, we find that the opportunity cost (of cell $O_2 D_1$) is $2 - 1 = +1$ dollar. This is the same value of opportunity cost (for cell $O_2 D_1$) that we found earlier by a direct observation of the net cost consequence associated with shifting 1 unit of goods into cell $O_2 D_1$. This equivalence holds for any empty cell, and we state again the general relationship:

$$\text{opportunity cost} = \text{implied cost} - \text{actual cost}$$

Although we have now succeeded in determining the opportunity cost of an empty cell by developing the concept of implied cost, it has been possible to do so only by first drawing a closed loop. The next logical question is: Can we somehow determine the implied cost of an empty cell without first drawing the closed loop? Should we find this to be possible, we would establish the main framework for the MODI method, for then we could subtract the actual cost of the empty cell from its calculated implied cost and thus determine its opportunity cost without first drawing the closed loop.

In this and the following paragraphs we shall develop a method for determining the implied costs of empty cells without drawing their respective loops. Let us refer back to the initial basic feasible solution of Table 9.16. In this program we have three occupied cells. In linear programming terms, this means that three of the four variables (that is, x_{11}, x_{12}, x_{22}) are basic variables. It will be recalled from the simplex method that the opportunity cost (represented by the $C_j - Z_j$ numbers in the net evaluation row) of any basic variable is zero. Similarly, it can be shown in the case of the transportation problem that the opportunity cost of each of the occupied cells (cells containing the basic variables) is zero. Now, if we assign a complete set of row numbers (to be placed at the extreme right-hand side of the table containing a given program) and a complete set of column numbers (to be placed at the bottom of the table) in such a way that the shipping cost per unit of *each* of the occupied cells equals the sum of its row and column numbers, we shall satisfy the condition that the opportunity cost of each occupied cell be zero.* Further, since the sum of the row and column numbers of any occupied cell equals the cost of that cell (a basic variable), the sum of the row and column numbers corresponding to each empty cell (nonbasic routes) gives the implied cost of that empty cell. The implied cost of any empty cell, therefore, is given by

$$\text{implied cost} = \text{row number} + \text{column number} = u_i + v_j$$

Thus, by the assignment of row and column numbers, we can calculate the implied cost of each empty cell without drawing a closed loop. We must now tackle the problem of assigning these row and column numbers.

For each occupied cell, we have to choose u_i (row number) and v_j (column number) such that c_{ij} (the actual shipping cost per unit in the occupied cell) equals the sum of u_i and v_j. For the occupied cell falling in row 1 and column 1, for example, u_1 and v_1 are chosen such that $c_{11} = u_1 + v_1$. Similarly, for cell O_1D_2, we must choose u_1 and v_2 such that $c_{12} = u_1 + v_2$. This process must be carried out for all the occupied cells. But it should be realized that although a basic feasible solution for a transportation problem consists of $m + n - 1$ variables (in other words, there are $m + n - 1$ occupied cells), we must assign $m + n$ values to obtain a complete set of row and column numbers. Hence, to determine all the row and column numbers, one arbitrary number, serving as either a row or a column number, must be chosen. Once one row number or column number has been chosen arbitrarily, the rest of the row and column numbers can be determined by the relationship $c_{ij} = u_i + v_j$. This relationship, as stated earlier, must hold for all the *occupied* cells. Insofar as any arbitrary number can be chosen to represent one of the u_i's or v_j's, we shall follow the practice of making u_1 take the value zero. This completes the description of the procedure for determining the row and column numbers.

The numbers for our example are given in Table 9.17. If we arbitrarily choose a value of zero for u_1, our next question is: What value must be given to v_1 so that $c_{11} = u_1 + v_1$ or $2 = 0 + v_1$? Obviously, v_1 must take a value

---

* The transportation problem, when represented in the simplex tableau format, consists of column vectors representing structural and other variables. In this problem, in each column vector representing a structural variable, two of the entries are 1 and the rest are 0. It is this special property of the transportation problem which makes it quite easy to test optimality by the MODI method. If the dual of the transportation problem is formed, the coefficients of all the dual variables will also be 1 or zero. Further, since for a 2×2 problem, the basic feasible solution has exactly three positive components in the primal, in the dual only three relationships need to be satisfied as equations. In the dual, while searching for the values of the dual variables, we assign values such that three constraints are satisfied as exact equalities, and these values must simultaneously satisfy the fourth "less than or equal to" constraint. These remarks can be extended to any $(m \times n)$ minimization problem. Their practical significance is this: A set of row numbers u_i and a set of column numbers v_j are chosen so that the opportunity cost of each cell is given by $u_i + v_j - c_{ij}$, where c_{ij} is the actual shipping cost per unit of the cell falling in ith row and jth column. Thus, if we choose u_i and v_j such that for all the occupied cells (basic routes) $c_{ij} = u_i + v_j$, we satisfy the requirement that the opportunity cost of each occupied cell is zero. For the empty cells (nonbasic routes), opportunity cost is given by $u_i + v_j - c_{ij}$.

Table 9.17

	Origin	Destination		Row Number
O_1	(900) [2]	(100) [2]		0
O_2		[1] (600) [2]		0
Column Number		2	2	

of 2. Next, we ask: What value must be given to v_2 so that $c_{12} = u_1 + v_2$ or $2 = 0 + v_2$? The value of v_2 must be 2. Again, what value must be given to u_2 so that $c_{22} = u_2 + v_2$ or $2 = u_2 + 2$? Obviously, $u_2 = 0$. By first assigning an arbitrary value to u_1 and then posing a series of questions, we have determined all the row and column numbers.

Let us now calculate the opportunity cost for the empty cell $O_2 D_1$. The opportunity cost of an empty cell, as stated earlier, is given by implied cost − actual cost, that is, by $(u_i + v_j) - c_{ij}$. For cell $O_2 D_1$, therefore, the opportunity cost is $u_2 + v_1 - c_{21} = 0 + 2 - 1 = +1$ dollar. The answer, of course, is the same as that obtained by the long method. Insofar as the opportunity cost of the empty cell $O_2 D_1$ is positive, this is not an optimal program and hence must be revised.

Before we revise the preceding program, let us summarize the role of the row and column numbers. Insofar as the row and column numbers are assigned in such a manner that the actual cost of every occupied cell equals the sum of its row and column numbers, the sum of the row and column numbers of each empty cell gives the implied cost of that empty cell (unused route). If the implied cost of the empty cell is less than its actual cost, this route should be left out of our shipping program. If, on the other hand,

Table 9.18

Implied Cost	Actual Cost	Action
$u_i + v_j >$	c_{ij}	A better program can be designed by including this cell in the solution.
$u_i + v_j =$	c_{ij}	Indifferent; however, an alternative program with the same total cost and including this cell can be designed.
$u_i + v_j <$	c_{ij}	Do not include this cell in the program.

the implied cost $(u_i + v_j)$ of an empty cell is more than its actual cost (c_{ij}), this route would be a candidate for inclusion in our next program. In summary, to evaluate and improve a given program in which the objective is to minimize a given function, the rules given in Table 9.18 apply. For a transportation problem in which the objective is to maximize a given function, the signs of the inequalities given in the Table 9.18 must be reversed to establish the guidelines for action.

Let us now return to our problem.

9.9.2 Design a New and Better Program

The last step in the MODI method is exactly the same as the corresponding step in the stepping-stone method. Having identified the empty cell to be included in the next program (the cell with the highest opportunity cost), we draw a closed loop for this cell. The new basic feasible solution is then derived by shifting into the empty cell the maximum possible number of units without violating the rim requirements. The second program is given in Table 9.19. To determine if the second program is an optimal program, we

Table 9.19

First Program — Second Program

must determine the opportunity cost of the empty cell $O_2 D_2$. This is illustrated in Table 9.20. From the table, we see that

$$\text{implied cost of cell } O_2 D_2 = u_2 + v_2 = -1 + 2 = +1$$
$$\text{actual cost of cell } O_2 D_2 = +2$$

Hence

$$\text{opportunity cost of empty cell } O_2 D_2 = \text{implied cost} - \text{actual cost}$$
$$= +1 - 2 = -1$$

Insofar as the opportunity cost of the only empty cell is nonpositive, no improvement in the present program is possible. This program, then, is the optimal program.

Table 9.20

Origin	Destination		Row Number
	D_1	D_2	
O_1	[2] 300	[2] 700	0
O_2	[1] 600	[2]	-1
Column Number	2	2	

We shall now apply the MODI method to the problem of Table 9.2.

STEP 1. Obtain an initial basic feasible solution.

As discussed earlier, an initial basic feasible solution for a given transportation problem may be obtained by following the northwest-corner rule, the least-cost method, or the penalty method. Table 9.21 reproduces the basic feasible solution of Table 9.4 (initial assignment by the least-cost method), which we shall take as a starting point for obtaining the optimal solution by MODI. Insofar as the number of occupied cells in this program equals $m + n - 1$, that is, $3 + 4 - 1 = 6$, this is indeed a basic feasible solution.

Table 9.21. Initial Assignment by Least-Cost Method

Origin	Destination				Total
	D_1	D_2	D_3	D_4	
O_1	[12] 10	[4]	[9] 25	[5] 20	55
O_2	[8]	[1] 20	[6] 25	[6]	45
O_3	[1] 30	[12]	[4]	[7]	30
Total	40	20	50	20	

STEP 2. Test for optimality (determine the opportunity costs of the empty cells).

To determine the opportunity costs of the empty cells by the MODI method, we must first determine the implied costs of the empty cells by assigning a complete set of row and column numbers. This is shown in Table 9.22. The *uncircled numbers* in the matrix represent the implied costs of the empty cells. Comparison of the implied and actual costs of each empty cell shows that only cell $O_2 D_1$ has a positive opportunity cost of $+1$ dollar. For cell $O_2 D_1$, opportunity cost = implied cost − actual cost = $9 - 8 = +1$. A similar calculation for cell $O_1 D_2$ shows that its opportunity cost is zero. The opportunity costs for the rest of the empty cells are negative.

Having identified the presence of positive opportunity cost, we know that this program is not an optimal program. Hence it must be revised to include that empty cell which has the highest opportunity cost (in this case cell $O_2 D_1$).

Table 9.22

Origin	Destination				Row Number
	D_1	D_2	D_3	D_4	
O_1	12 (10)	4 / 4	9 (25)	5 (20)	0
O_2	8 / 9	1 (20)	6 (25)	6 / 2	−3
O_3	1 (30)	12 / −7	4 / −2	7 / −6	−11
Column Number	12	4	9	5	

STEP 3. Design a new and better program.

The revision of any program is guided by a closed loop drawn for the *most favorable* empty cell that is to be included in the next program. Here the loop for cell $O_2 D_1$ is drawn and the program is revised in exactly the same manner as shown in Table 9.14. The revised program is then tested for optimality (by assigning values to row numbers, u_i, and column numbers, v_j) as shown in Table 9.23. A comparison of the uncircled numbers (representing

Table 9.23

Origin	Destination				Row Number
	D_1	D_2	D_3	D_4	
O_1	12 11	4 4	9 (35)	5 (20)	0
O_2	8 (10)	1 (20)	6 (15)	6 2	−3
O_3	1 (30)	12 −6	4 −1	7 −5	−10
Column Number	11	4	9	5	

the implied costs) in the empty cells and the respective actual costs shows that no empty cell has a positive opportunity cost.* Hence this is an optimal solution.†

9.10 PROCEDURE SUMMARY FOR THE MODIFIED DISTRIBUTION METHOD (MINIMIZATION CASE)

STEP 1. Obtain a basic feasible solution.

An initial basic feasible solution for a given transportation problem may be obtained by the northwest-corner rule, the least-cost method, or the penalty method.

Test for step 1. A basic feasible solution must include shipments covering $m + n - 1$ cells. That is, the number of occupied cells (basic variables) is 1 less than the number of rows and columns in the transportation matrix.

If the number of occupied cells in the initial solution is more than $m + n - 1$, there is a computational error, which can be corrected easily by rechecking the data. If the number of occupied cells is less than $m + n - 1$,

---

* The fact that empty cell $O_1 D_2$ has an opportunity cost of zero means that an alternative program which will include cell $O_1 D_2$ and have the same total shipping cost as this program can be designed.

† The implied cost of an empty cell gives an indication of the *cost sensitivity* of that unused route. For example, the cells $O_3 D_2$, $O_3 D_3$, and $O_3 D_4$ in Table 9.23 are insensitive to transportation costs, whereas $O_1 D_1$, $O_1 D_2$, and $O_2 D_4$ are not.

this is a degenerate solution. To resolve degeneracy, add one or more epsilons to some "suitable" empty cells so that the number of occupied cells becomes equal to $m + n - 1$.*

STEP 2. Test for optimality (determine the opportunity costs of the empty cells).

a. *Determine a complete set of row and column numbers (values).* When, in a given program, the number of occupied cells equals $m + n - 1$, proceed to assign row and column numbers (values) in such a manner that, for each occupied cell, the relationship $c_{ij} = u_i + v_j$ holds. To start, a value of zero can be assigned to any row. The rest of the row and column numbers can then be determined by making sure that, for *each occupied cell*, $c_{ij} = u_i + v_j$. In other words, for each occupied cell, the actual shipping cost per unit should equal the sum of its row and column values.

b. *Calculate the implied costs of the empty cells.* Once all the row and column values have been assigned, the implied cost of a given empty cell can be calculated as follows:

$$\text{implied cost} = \text{row value} + \text{column value}$$

c. *Determine the opportunity costs of the empty cells.* The opportunity cost of an empty cell is determined by subtracting the actual cost of the empty cell from its implied cost. In other words, opportunity cost, for each cell, is given by

$$\text{opportunity cost} = (u_i + v_j) - c_{ij}$$

If the opportunity costs of all the empty cells are nonpositive, an optimal solution has been obtained. If, on the other hand, even a single cell has a positive opportunity cost, a better program can be designed. Thus step 2 serves as a test for optimality.

STEP 3. Design a new and better program.

Design a new program such that the empty cell that has the largest opportunity cost (in the program to be revised) is included in the solution. This is accomplished in the following manner:

a. Draw a loop of horizontal and vertical arrows in such a manner that it starts from the empty cell to be filled, passes to an occupied cell in the same row or column as the empty cell, and then, making a series of alternate horizontal and vertical turns through occupied cells, returns to the original empty cell.

b. Place a plus sign ($+$) in the empty cell to be filled. Then, alternately, place minus signs ($-$) and plus signs ($+$) at the beginnings and ends of the connecting links of the loop.

---

* Degeneracy in transportation problems is discussed in Section 9.14. Epsilon is a *very* small number that can be ignored when the optimal solution has been obtained.

c. Examine those occupied cells in which the minus signs have been placed. Of these, the cell that has the least number of units is vacated by transferring these units to the empty cell. This is accomplished by adding the same amount to all cells that have plus signs and subtracting it from all cells that have minus signs. The improved program should have the same number of occupied cells as the preceding program. If the number of occupied cells in the improved program is less than that of the preceding program, the problem becomes degenerate. In such a case, add epsilon(s) to some recently vacated cell(s) such that the number of occupied cells again equals $m + n - 1$.*

STEP 4

Repeat steps 2 and 3 until a program is achieved in which each empty cell has an opportunity-cost value that is either zero or negative. This program will be the optimal program.

9.11 THE MODIFIED DISTRIBUTION METHOD (MAXIMIZATION CASE)

Except for one transformation, a transportation problem in which the objective is to maximize a given function can also be solved by the MODI algorithm. The transformation is made by subtracting all the c_{ij}'s from the highest c_{ij} (profit) of the given transportation matrix. The transformed c_{ij}'s give us the relative costs, and the problem then becomes a minimization problem. Once an optimal solution to this transformed minimization problem has been found, the value of the objective function can be calculated by inserting the original values of the c_{ij}'s for those routes which form the basis (occupied cells) in the optimal solution.

9.12 BALANCING THE GIVEN TRANSPORTATION PROBLEM

To solve a given transportation problem by the step-by-step procedure given in Section 9.10, we must establish equality between the total capacities of the origins and the total requirements of the destinations. Three cases can arise.

CASE 1. $\sum b_i = \sum d_j$

In this case, the total capacity of the origins equals the total requirement of the destinations. The problem can be arranged in the form of a matrix,

---

* See Section 9.14 for a discussion of degeneracy in transportation problems.

along with the relevant cost data, and the transportation algorithm may be applied directly to obtain a solution.

CASE 2: $\sum b_i > \sum d_j$

In this case, the total capacity of the origins exceeds the total requirement of the destinations. A "dummy" destination can be added to the matrix to absorb the excess capacity. The cost of shipping from each origin to this dummy destination is assumed to be zero. The addition of a dummy destination establishes equality between the total origin capacities and total destination requirements. The problem is then amenable to solution by the transportation algorithm.

EXAMPLE 9.1

Table 9.24 gives both the unbalanced and balanced forms of a transportation problem in which the total given capacity of the origins exceeds the total given requirement of the destinations ($\sum b_i > \sum d_j$). The balanced problem can be solved by the transportation method, and the optimal solution will identify the particular origin at which the excess capacity should be left idle.

CASE 3: $\sum b_i < \sum d_j$

In this case, the total capacity of the origins is less than the total requirement of the destinations. A dummy origin can be added to the transportation matrix to meet the excess demand. The cost of shipping from the dummy origin to each destination is assumed to be zero. The addition of a dummy origin in this case establishes equality between the total capacity of the origins and the total requirement of the destinations.*

EXAMPLE 9.2

Table 9.25 gives both the unbalanced and balanced forms of a transportation problem in which the total given capacity of the origins is less than the total requirement of the destinations ($\sum b_i < \sum d_j$). The balanced problem can be solved by the transportation method, and the optimal solution will identify the particular destination whose requirement cannot be fully satisfied.

In the initial assignment for a transportation problem that has been balanced by the addition of a dummy origin or a dummy destination, only the last necessary allocations should be made to the dummy cells. This procedure, in general, will result in fewer iterations.

---

* The reader will observe that in a transportation problem, the role of the dummy column or dummy row that contains dummy variables is parallel to the role of the slack variables in the general linear programming problems illustrated previously.

Table 9.24

Unbalanced Form

	D_1	D_2	D_3	Origin Capacity
O_1	5	3	2	200
O_2	6	4	1	400
Destination Requirement	200	200	150	

Balanced Form

	D_1	D_2	D_3	Dummy	Origin Capacity
O_1	5	3	2	0	200
O_2	6	4	1	0	400
Destination Requirement	200	200	150	50	

245

Table 9.25

Unbalanced Form

	D_1	D_2	D_3	Origin Capacity
O_1	5	3	2	200
O_2	6	4	1	400
Destination Requirement	300	200	150	

Balanced Form

	D_1	D_2	D_3	Origin Capacity
O_1	5	3	2	200
O_2	6	4	1	400
Dummy	0	0	0	50
Destination Requirement	300	200	150	

9.13 COMPARISON OF THE SIMPLEX AND TRANSPORTATION METHODS

Some important observations can be made regarding parallels in the general transportation model and the simplex method. First, the role of the dummy variables in the transportation problem is similar to the role of the slack variables in the general linear programming problem. Second, the occupied cells and empty cells of the transportation program correspond, respectively, to the basic variables and nonbasic variables of the simplex tableau.

Third, the revision of a given transportation program is parallel to the process of obtaining a new basis in the simplex method. Let us explain this point further. A given transportation program, it will be recalled, is improved by filling one empty cell (the one that has the highest opportunity cost) at a time. In this process, all the units from at least one currently occupied cell are removed. Thus a new cell is filled and becomes an occupied cell, and at least one of the previously occupied cells joins the category of empty cells. The total number of occupied cells, therefore, can either remain constant (*only one* previously occupied cell becomes an empty cell) or decrease (*more than one* of the previously occupied cells become empty cells) from one program to the next.

If the number of occupied cells remains the same from one program to the next, the process is similar to a simplex iteration in which one nonbasic variable is introduced into the solution to remove one of the basic variables currently in the solution. Of course, in this case we obtain a new basic feasible solution. If, on the other hand, the process of filling one empty cell results in the simultaneous vacating of two or more of the currently occupied cells, the transportation problem becomes degenerate. The latter situation, as the reader will observe, is parallel to the simplex iteration in which the introduction of one new (nonbasic) variable removes, simultaneously, two or more of the current basic variables.

9.14 DEGENERACY IN TRANSPORTATION PROBLEMS

It was established earlier that a basic feasible solution for a transportation problem consists of $m + n - 1$ basic variables. This means that the number of occupied cells in a given transportation program is 1 less than the number of rows and columns in the transportation matrix. Whenever the number of occupied cells is less than $m + n - 1$, the transportation problem is said to be *degenerate*.

Degeneracy in transportation problems can develop in two ways. First, the problem may become degenerate when the initial program is designed via one of the initial-assignment methods discussed earlier. Second, the transportation problem may become degenerate during the solution stages. This happens when the inclusion of the most favorable empty cell (the cell having the highest opportunity cost) results in the simultaneous vacating of two or more of the currently occupied cells. To resolve degeneracy in both cases, we can allocate an extremely small amount of goods (close to zero) to one or more of the empty cells,* so that the number of occupied cells becomes $m + n - 1$. The cell containing this extremely small allocation is, of course, considered to be an occupied cell.

In linear programming literature, this extremely small amount is usually denoted by the Greek letter ϵ (epsilon). The amount ϵ is assumed to be so small that its addition to or subtraction from a given number does not change that number. For example, $50 + \epsilon = 50$, and $200 - \epsilon = 200$. Of course, if ϵ is subtracted from itself, the result is assumed to be zero; that is, $\epsilon - \epsilon = 0$.

Once the number of occupied cells is $m + n - 1$, the transportation method can be applied in a straightforward manner. If, in the optimal solution, an occupied cell has ϵ units, it implies that no units are shipped through that route.

9.15 ALTERNATIVE OPTIMAL SOLUTIONS TO TRANSPORTATION PROBLEMS

An optimal solution to a given transportation problem is not always a unique solution. The existence of more than one optimal solution for a transportation problem can be determined by examining the opportunity costs of the empty cells in the optimal program designed by following the transportation algorithm. If any empty cell has an opportunity cost of zero in the optimal program, another optimal program with the same total shipping cost as the first can always be designed. The second optimal program is obtained by revising the first program so as to include the zero-opportunity-cost cell.

Once the existence of two alternative optimal programs is established, an infinite number of other alternative optimal programs can be derived. The following relationship governs the derivation of these alternative programs:

$$\text{derived program} = d\mathbf{A} + (1 - d)\mathbf{B}$$

where

$\mathbf{A}$ = matrix representing first optimal program

$\mathbf{B}$ = matrix representing second optimal program

d = any positive fraction less than 1.

---

* This extremely small amount may be allocated to any empty cell subject to the condition that this will make possible the determination of a unique set of row and column numbers.

Insofar as we can let d be any positive fraction, it is obvious that an infinite number of derived solutions can be obtained as long as two alternative optimal solutions can be identified.

In terms of practical significance, the possibility of designing alternative solutions gives valuable flexibility to the manager. It should also be realized that an examination of the opportunity costs of the empty cells (or the optimal program) enables us to identify solutions in descending order of preference in terms of total shipping cost.

REFERENCES

See references at the end of Chapter 7.

REVIEW QUESTIONS AND PROBLEMS

9.1. In what sense is the transportation model a special case of the general linear programming model?

9.2. What is a balanced transportation problem? Describe the approach of the transportation method.

9.3. What are the three methods of making initial assignments in a transportation problem?

9.4. Explain the rationale of the MODI method of testing optimality. How does MODI method differ from the stepping-stone method?

9.5. What are rim requirements? What role do they play in revising a given transportation program?

9.6. What are the row and column numbers? Relate the concept of dual variables to the row and column numbers.

9.7. Distinguish between the terms "opportunity cost" and "implied cost" of an empty cell.

9.8. Rakesh Gupta Manufacturing Company manufactures a very expensive executive desk. The demand for the desks is so high that the company operates four plants at full production. The company supplies four wholesalers located in various parts of the country. The cost of shipping 1 unit from a particular plant to a specific warehouse is shown in the relevant upper right-hand corner of the cell located at the intersection of the plant row and warehouse column (see Table 9.26).
(a) Design an initial shipping program by the northwest-corner rule.
(b) Find the optimal shipping program.

9.9. The Smart Shopper Co. wishes to ship a specific brand of shirts to its four stores. Warehouse capacities, expected store-demands levels, and costs to ship one unit from each warehouse to each store are given in Table 9.27. Utilize the MODI method to design an optimal shipping schedule. Suppose that store B goes out of business; this means that there is an excess capacity of 300 units. Find the minimum cost transportation schedule.

Table 9.26

Plant	Boston	Chicago	Dallas	Kansas City	Seattle	Plant Capacity
Newark	1	2	6	2	3	800
Atlanta	3	4	5	8	1	600
Cleveland	3	1	1	2	6	200
Denver	4	7	3	5	4	400
Warehouse Demand	400	100	700	300	500	

Table 9.27

Warehouse	A	B	C	D	Warehouse Capacity
1	8	6	3	4	300
2	2	5	4	7	400
3	6	8	3	3	600
Store Demand	250	300	500	250	

9.10. (a) What restrictions make the transportation problem a special case of the general linear programming model?

(b) What happens if the initial assignment in a transportation problem gives less than $m + n - 1$ occupied cells, where m is the number of origins and n is the number of destinations? How can such a problem be solved?

9.11. Solve the transportation problem shown in Table 9.28 (use the northwest-corner rule to design an initial program).

Table 9.28

Factory	Warehouse					Capacity Restrictions
	A	B	C	D	E	
X	37	27	28	34	31	100
Y	29	31	32	27	29	125
Z	33	26	35	30	30	150
Warehouse Requirement	50	60	70	80	90	

Table 9.29

Plant	Warehouse			Plant Capacity
	A	B	C	
X	6	7	5	225
Y	11	3	4	150
Z	8	4	3	375
Warehouse Requirement	100	200	450	

9.12. The Ace Company has to ship a homogeneous commodity from three plants to three warehouses. The transportation cost per unit from each factory to each warehouse, the requirements of each warehouse, and the capacity of each plant are shown in Table 9.29.

(a) Find the minimum-cost transportation program.

(b) Find the optimal program if the capacity of plant X is increased to 400.

9.13. According to projected demand, there will be a surplus of 8, 7, 9, and 10 tank-loads of gasoline in towns A, B, C, and D, respectively, by the beginning of next month. Towns K, L, M, and N are expected to have a deficit of 6, 7, 9, and 8 tank loads of gasoline, respectively. The dollar cost of moving tankloads from each surplus town to each deficit town is given in Table 9.30.

Table 9.30

Surplus Town	Deficit Town			
	K	L	M	N
A	27	24	25	22
B	17	14	21	38
C	16	16	28	17
D	33	27	32	26

(a) Determine the optimum allocation schedule.

(b) What is the cost of the optimal program?

9.14. Bottles of wines from wine-producing regions X, Y, and Z are to be shipped to six market areas, A, B, C, D, E, and F. The production of wine in regions X, Y, and Z is 7,000, 6,750, and 6,000 bottles, respectively. The demand for the wine in market areas A, B, C, D, E, and F is 2,500, 2,400, 1,850, 3,200, 2,700, and 3,400 bottles, respectively. The cost of shipping 100 bottles of wine from each region to each market is shown in Table 9.31.

Table 9.31

Region	Market Area					
	A	B	C	D	E	F
X	24	16	26	25	17.5	20
Y	19	22	24	15	17.5	27
Z	23	17.5	27	19	20	25

(a) Find the optimal shipping program.

(b) What is the cost of the optimal program?

10

The Assignment Model

10.1 INTRODUCTION

The assignment model deals with a special class of linear programming problems in which the objective is to assign a number of "origins" to the same number of "destinations" at a minimum total cost.* The assignment is to be made on a one-to-one basis. That is, *each origin can associate with one and only one destination.* This feature implies the existence of two specific characteristics in a linear programming problem, which, when present, give rise to an assignment problem. First, the payoff matrix for the given problem is a square matrix. Second, the optimal solution (or any solution within the given constraints) for the problem is such that there can be one and only one assignment in a given row or column of the payoff matrix.

Payoffs for each assignment are assumed to be known and independent of each other. With information about the number of origins and destinations and the payoffs associated with each available assignment, the assignment model is used to choose the strategy that maximizes or minimizes the total payoff, depending upon whether the particular payoff represents a gain or a loss to the decision-maker.

---

* Of course, if the payoff is of the profit variety, the objective is to maximize total payoff.

10.2   A SIMPLE ASSIGNMENT PROBLEM

Let us illustrate an extremely simple assignment problem by considering the assignment of three jobs, O_1, O_2, and O_3 to three machines, D_1, D_2, and D_3. The problem states that any one of the jobs can be processed completely with any one of the machines. Further, the cost of processing the ith job ($i = 1, 2, 3$) with the jth machine ($j = 1, 2, 3$) is known. The objective, therefore, is to assign these jobs to the machines in a manner that will minimize the total cost of processing all the jobs. The relevant cost data are given in Table 10.1. A quick visual inspection of this simple problem reveals that the minimum total cost assignment will require that jobs O_1, O_2, and O_3 be assigned, respectively, to machines D_1, D_2, and D_3.

Table 10.1

Job	Machine		
	D_1	D_2	D_3
O_1	10	15	20
O_2	19	12	16
O_3	12	14	11

The total cost of the optimal assignment can be obtained by multiplying the cost of each assigned (occupied) cell by 1, multiplying the cost of each unassigned (empty) cell by 0, and then adding the products. Thus the total cost of the optimal assignments for this problem is

$$1(10) + 0(19) + 0(12) + 0(15) + 1(12) + 0(14)$$
$$+ 0(20) + 0(16) + 1(11) = \$33$$

This suggests that the optimal-assignment matrix can be represented as in Table 10.2. In other words, we can think of the assignment problem as a problem in making proper "matches" between the origins and the destinations. A value of 1 is allocated to those cells* for which a match has been made; a value of 0 is given to all other cells.

There are various methods for making these matches. First, as will be shown in Section 10.4, we can use the transportation model for solving the assignment problem. Second, provided that the problem is of small dimensions, we can identify the optimal assignment by enumerating and examining all the possible assignments (see Section 10.5). Third, we can use the assignment model for solving such problems (see Section 10.7). Of these,

---

* This identifies a complete utilization and satisfaction of the capacity and requirement of the particular row and column in which such a cell falls.

Table 10.2. Optimal-Assignment Matrix

Job	Machine		
	D_1	D_2	D_3
O_1	1		
O_2		1	
O_3			1

the assignment model is the most efficient method of attack for obtaining the optimal assignment.

We have constructed and solved the problem of Table 10.1 with the objective of giving the reader an intuitive understanding of the assignment problem. Note that the optimal-assignment matrix has one and only one assignment in each row and each column. Another way of saying the same thing is that the sum of assignments for each row and column in the optimal solution to the assignment problem must be 1. This requirement, along with other components of a complete statement of the assignment problem, is presented in the next section.

10.3 THE ASSIGNMENT PROBLEM AS A SPECIAL CASE OF THE TRANSPORTATION PROBLEM

It was mentioned earlier that the assignment problem is a special case of the general linear programming problem. As a matter of fact, the assignment problem is a special case of the transportation problem, which, in turn, is itself a special case of the general linear programming problem. This will become clear as we consider the 3×3 transportation problem in Table 10.3. It will be recalled that a transportation problem of this type can be stated as:

Minimize
$$F(X) = \sum_{j=1}^{3} \sum_{i=1}^{3} c_{ij} x_{ij}$$

subject to the constraints

$$\sum_{j=1}^{3} x_{ij} = b_i \qquad i = 1, 2, 3$$

$$\sum_{i=1}^{3} x_{ij} = d_j \qquad j = 1, 2, 3$$

and

$$x_{ij} \geqslant 0 \qquad \begin{array}{l} i = 1, 2, 3 \\ j = 1, 2, 3 \end{array}$$

Table 10.3

Origin	Destination			Origin Capacity per Time Period
	D_1	D_2	D_3	
O_1	c_{11}	c_{12}	c_{13}	b_1
O_2	c_{21}	c_{22}	c_{23}	b_2
O_3	c_{31}	c_{32}	c_{33}	b_3
Destination Requirement per Time Period	d_1	d_2	d_3	

In other words, this transportation problem calls for the determination of the x_{ij}'s such that the objective function is minimized subject to the given constraints. Now, let us suppose that each $b_i = 1$ and each $d_j = 1$. Impose, further, the restriction that $x_{ij} = 0$ or 1. Then the transportation problem reduces to the following form:

Minimize
$$F(X) = \sum_{j=1}^{3} \sum_{i=1}^{3} c_{ij} x_{ij}$$

subject to the constraints

$$\left. \begin{array}{ll} \sum_{j=1}^{3} x_{ij} = 1 & i = 1, 2, 3 \\ \\ \sum_{i=1}^{3} x_{ij} = 1 & j = 1, 2, 3 \end{array} \right\} \begin{array}{l} \text{structural} \\ \text{constraints} \end{array}$$

and

$$x_{ij} = 0 \quad \text{or} \quad 1$$

The mathematical problem given above corresponds exactly to the descriptive statement of the assignment problem of Table 10.1. Therefore, we see that the assignment problem is, indeed, a special case of the transportation problem.

10.4 SOLVING AN ASSIGNMENT PROBLEM BY THE TRANSPORTATION ALGORITHM

Insofar as the assignment problem is a special case of the transportation problem, we should be able to solve any assignment problem by application of the transportation algorithm. We shall illustrate this by considering the assignment problem in Table 10.4. The number within each cell represents the cost (c_{ij}) of processing the ith job with the jth machine.

In view of our previous discussion, we place this problem in a transportation format in Table 10.5.

Let us use the MODI method to solve this problem. The first step, of course, is to obtain a basic feasible solution by making an initial assignment. Following the northwest-corner rule, we obtain the initial program given in

Table 10.4

Job	Machine		
	D_1	D_2	D_3
O_1	20	27	30
O_2	10	18	16
O_3	14	16	12

Table 10.5

Job	Machine			Job Requirement
	D_1	D_2	D_3	
O_1	20	27	30	1
O_2	10	18	16	1
O_3	14	16	12	1
Machine Capacity	1	1	1	

Table 10.6a. The number of occupied cells in this program is 3. However, we need $m + n - 1$, that is, 5, occupied cells to obtain a basic feasible solution.*

We add ϵ's to cells $O_1 D_2$ and $O_2 D_3$, thus obtaining the basic feasible solution given in Table 10.6b.† An assignment of row and column numbers to this program shows that the empty cell $O_2 D_1$ has a positive opportunity cost, since the implied cost of cell $O_2 D_1$ is 11, whereas its actual cost is 10 (see Table 10.6b). Since the program of Table 10.6b does not represent an optimal program (the opportunity cost of cell $O_2 D_1$ is positive), we revise this program to include cell $O_2 D_1$ (see Table 10.6c).

The second program, shown in Table 10.6d, can be tested for optimality after it is made a basic feasible solution by the addition of ϵ to either cell $O_2 D_2$ or cell $O_3 D_2$ and a set of row and column numbers is obtained. This is done in Table 10.6e. A visual inspection of Table 10.6e shows that the opportunity costs of all the empty cells are nonpositive; hence our second program is the optimal assignment.

Assign job O_1 to machine D_2.
Assign job O_2 to machine D_1.
Assign job O_3 to machine D_3.

The total cost of this optimal assignment is $27 + 10 + 12 = \$49$.

Table 10.6a. Initial Program by Northwest-Corner Rule (Degenerate Solution)

	D_1	D_2	D_3	Total
O_1	20 ①	27	30	1
O_2	10	18 ①	16	1
O_3	14	16	12 ①	1
Total	1	1	1	

---

* Insofar as any assignment problem must have a square payoff matrix (say $n \times n$), a basic feasible solution should have $n + n - 1$ occupied cells. But, owing to the structural constraints of the assignment problem, any solution of such a problem cannot have more than n assignments (i.e., n occupied cells). Hence *the assignment problem is inherently degenerate.*

† We add ϵ's to that empty cell(s) which permits (permit) the use of the MODI method to compute row and column numbers.

Table 10.6b. A Basic Feasible Solution

	D_1	D_2	D_3	Row Number
O_1	[20] ①	[27] ⓔ	[30]	0
O_2	[10]	[18] ①	[16] ⓔ	−9
O_3	[14]	[16]	[12] ①	−13
Column Number	20	27	25	

Table 10.6c. First Program

	D_1	D_2	D_3	Total
O_1	[20] ①	[27] ⓔ	[30]	1
O_2	[10]	[18] ①	[16] ⓔ	1
O_3	[14]	[16]	[12] ①	1
Total	1	1	1	

By solving the assignment problem of Table 10.4 by the transportation technique, we have added little to our knowledge, since the transportation method was adequately covered in Chapter 9. The effort has not been completely wasted, however, for we have shown that any assignment problem can be solved by the transportation technique.

Table 10.6d. Second Program

	D_1	D_2	D_3	Total
O_1	[20]	[27] ①	[30]	1
O_2	[10] ①	[18]	[16] ⓔ	1
O_3	[14]	[16]	[12] ①	1
Total	1	1	1	

Table 10.6e. Second and Optimal Program

	D_1	D_2	D_3	Row Number
O_1	[20]	[27] ①	[30]	0
O_2	[10] ①	[18] ⓔ	[16] ⓔ	-9
O_3	[14]	[16]	[12] ①	-13
Column Number	19	27	25	

10.5 SOLVING AN ASSIGNMENT PROBLEM BY ENUMERATION

The assignment problem, if time and money are assumed to be unlimited, can also be solved by first enumerating all possible assignments and then choosing the least-cost assignment. For example, there are six possible assignments for the problem of Table 10.4.* These assignments, along with their respective total costs, are listed in Table 10.7. Obviously, assignment program 3, with a total cost of $49, is the least-cost or optimal assignment.

Needless to say, we are not seriously advocating the solution of assignment problems by enumeration. The reader has only to think of a 10×10 assignment problem, not to speak of larger dimensions, to realize that the solution by enumeration is impractical.† However, this way of looking at the assignment method shows us the significance of more efficient methods of attack.

Table 10.7

	Assignment Program	Total Cost ($)
1	$O_1 D_1, O_2 D_2, O_3 D_3$	$20 + 18 + 12 = 50$
2	$O_1 D_1, O_2 D_3, O_3 D_2$	$20 + 16 + 16 = 52$
3	$O_1 D_2, O_2 D_1, O_3 D_3$	$27 + 10 + 12 = 49$
4	$O_1 D_2, O_2 D_3, O_3 D_1$	$27 + 16 + 14 = 57$
5	$O_1 D_3, O_2 D_2, O_3 D_1$	$30 + 18 + 14 = 62$
6	$O_1 D_3, O_2 D_1, O_3 D_2$	$30 + 10 + 16 = 56$

10.6 APPROACH OF THE ASSIGNMENT MODEL

We are now aware that it is at least theoretically possible to solve a given assignment problem by application of the transportation technique. However, a much more efficient method of solving such problems is available. This method of solving assignment problems, known as the *Hungarian method*, will be referred to as the *assignment model* or the *assignment method*.

The assignment method consists of three basic steps. The first step involves the derivation of a "total-opportunity-cost" matrix from the given payoff matrix of the problem. This is done, as we shall illustrate in Section 10.7, by (1) subtracting the lowest number of each column of the given payoff matrix from all the other numbers in its column, and (2) subtracting the lowest number of each row of the matrix obtained in (1) from all the other numbers in its row. The *total-opportunity-cost matrix* thus derived will have at least one zero in each row and column. Any cell that has an entry of zero in the total-opportunity-cost matrix is considered to be a candidate for assignment. The significance of the total-opportunity-cost matrix is that it presents some possible assignment alternatives in which the opportunity costs of some or all assignments may be zero.

---

* For an assignment problem that has an $n \times n$ payoff matrix, the number of possible assignments equals $n!$ Thus in this case we have 3!, or 6, possible assignments.

† There are 10! or 3,628,800, possible assignments in a 10×10 assignment problem.

The purpose of the second step is to determine whether an optimal assignment, guided by the total-opportunity-cost matrix derived in step 1, can be made. This is accomplished, as we shall see in the next section, by a simple test. If the test shows that an optimal assignment (with a total opportunity cost of zero) can be made, the problem is solved.* On the other hand, if an optimal assignment cannot be made, we proceed to step 3.

The purpose of the third step is to revise the current total-opportunity-cost matrix to derive some better assignments. The procedure by which this is accomplished either redistributes the zeros of the current total-opportunity-cost matrix or creates one or more new zero cells. The result is another total-opportunity-cost matrix which enables us to find a less costly assignment. In other words, the result of step 3 brings us back to the beginning of step 2, and we again search for an optimal solution. Thus steps 2 and 3 are repeated as many times as are necessary to find an optimal solution that has a total opportunity cost of zero.

The remaining sections of this chapter are devoted to illustrating the development and application of the assignment method.

10.7 DEVELOPMENT OF THE ASSIGNMENT METHOD

In this section we shall present an intuitive rationale for the various steps of the assignment method. A brief description of the mathematical foundation of the assignment method will be given in Section 10.11.

Let us consider again the assignment problem of Table 10.4. The relevant cost data are reproduced in Table 10.8.

Table 10.8

	Machine		
Job	D_1	D_2	D_3
O_1	20	27	30
O_2	10	18	16
O_3	14	16	12

STEP 1. Determine the total-opportunity-cost matrix.

Opportunity cost is the cost involved in *not* following the best course of action. With respect to a matrix, we can consider two types of opportunity costs:

---

* Insofar as the assignment problem involves decision making under certainty, the total opportunity cost of the optimal solution must be zero.

1. The *column opportunity cost*, which reflects the relative efficiency of a given column with respect to rows.
2. The *row opportunity cost*, which reflects the relative efficiency of a given row with respect to columns.

If we examine column D_1, we observe that machine D_1 is most efficient with respect to job O_2; hence the opportunity cost of matching D_1 with O_2 is zero ($10 - 10 = 0$). On the other hand, if we were to match D_1 with O_1, it will involve an opportunity cost of \$10 (i.e., $20 - 10 = 10$). Similarly, the opportunity cost of assigning O_3 to D_1 is \$4 (i.e., $14 - 10 = 4$). By similar argument, it is possible to determine the column opportunity costs for columns D_2 and D_3. It should be clear by now that the column opportunity cost of a matrix can be determined by subtracting the lowest number, in *each* column, from all the numbers in that column. For the assignment problem of Table 10.8, the column-opportunity-cost matrix is shown in Table 10.9.

Table 10.9

	D_1	D_2	D_3
O_1	10	11	18
O_2	0	2	4
O_3	4	0	0

The column-opportunity-cost matrix was derived by determining the relative efficiency of a given column (machine) with respect to all the rows (jobs). It should now be observed that we can also analyze alternatives by examining the relative efficiency of a given row (job) with respect to all the columns (machines). For example, consider job O_1 with respect to the three machines. It is obvious that the row opportunity cost of assigning O_1 to D_1 is zero ($20 - 20 = 0$). However, the row opportunity cost of assigning O_1 to D_2 is \$7 (i.e., $27 - 20 = 7$); and the row opportunity cost of assigning O_1 to D_3 is \$10 (i.e., $30 - 20 = 10$). By similar argument, it is possible to determine the row opportunity costs of rows O_2 and O_3. It is clear that the row opportunity cost of a matrix can be determined by subtracting the lowest number in *each* row from all the numbers in that row. For the assignment problem of Table 10.8, the row-opportunity-cost matrix is shown in Table 10.10. Thus, in an assignment problem, any match between an origin and a destination gives rise to two types of opportunity costs. One is the opportunity cost with respect to the lowest payoff in the column to which the assignment cell belongs (we have called it column opportunity cost). The other is the opportunity cost with respect to the lowest payoff in the row to which the assignment cell belongs (we have called it row opportunity cost).

Table 10.10

	D_1	D_2	D_3
O_1	0	7	10
O_2	0	8	6
O_3	2	4	0

It should be emphasized that total opportunity cost is a *relative* concept. It measures *relative* efficiencies for the entire payoff matrix rather than relative efficiencies with respect only to columns or rows. It is possible to determine the total-opportunity-cost matrix directly from the original payoff matrix by performing two operations in sequence. First, subtract the lowest number, in each column, from all the numbers; then take the resultant matrix and subtract, in each row, the lowest number from all the numbers in that row. What we obtain is the total-opportunity-cost matrix, as shown in Table 10.11.

Table 10.11. Derivation of the Total-Opportunity-Cost Matrix

Original Cost Matrix				Column-Opportunity-Cost Matrix				Total-Opportunity-Cost Matrix		
	D_1	D_2	D_3		D_1	D_2	D_3		D_1	D_2	D_3
O_1	20	27	30	O_1	10	11	18	O_1	0	1	8
O_2	10	18	16	O_2	0	2	4	O_2	0	2	4
O_3	14	16	12	O_3	4	0	0	O_3	4	0	0

Let us summarize the two operations required in calculating the total-opportunity-cost matrix for any assignment problem:

a. Subtract the lowest entry in each column of the given payoff matrix from all the entries in its column.
b. Subtract the lowest entry in each row of the matrix obtained in (a) from all the numbers in its row.

STEP 2. Test for optimality (determine whether an optimal assignment can be made).

An optimal assignment in this problem can always be made if we can locate three zero cells* in the total-opportunity-cost matrix (see Table 10.11) such that a complete assignment to these cells can be made with a total opportunity cost of zero. This can happen only when no two such zero cells

---

* For an $n \times n$ assignment problem, we must locate n such zero cells. Such zeros, no two (or more) of which lie in the same row or column, are said to form a set of independent zeros.

are in the same row or column, regardless of the number of zero cells in the total-opportunity-cost matrix. Thus, of the four zero cells in the total-opportunity-cost matrix of Table 10.11, only two zero cells are useful for the purpose of obtaining an optimal assignment. On the other hand, if we had an additional zero, say in cell O_1D_2, we could locate a set of three zero cells such that an optimal assignment (with a total opportunity cost of zero) could be made.

Based on the above discussion, a simple test has been devised to determine whether an optimal assignment can be made. It consists in drawing and counting the *minimum* number of horizontal and vertical (*not* diagonal) lines necessary to cover all the zero cells in the total-opportunity-cost matrix. If the number of lines equals the number of rows (or the number of columns) of the given payoff matrix, an optimal assignment can be made, and the problem is solved. On the other hand, if the minimum number of lines needed to cover all the zero cells is less than the number of rows, or columns, an optimal assignment cannot be made, and it is necessary to construct a revised total-opportunity-cost matrix.

An application of this test to the total-opportunity-cost matrix of Table 10.11 shows that an optimal assignment cannot be made at this stage. It takes only two lines (row O_3 and column D_1) to cover all the zero cells in the total-opportunity-cost matrix (see Table 10.12a), whereas the number of rows is 3. Hence a revised total-opportunity-cost matrix, which will lead us toward an optimal assignment, must be obtained.

STEP 3. Revise the current total-opportunity-cost matrix.

After the application of the test of step 2, the cells of the total-opportunity-cost matrix can be classified into two categories:

1. The *covered cells*, which have been covered by the lines.
2. The *uncovered cells*, which have not been covered by the lines.

The fact that an optimal assignment cannot be made may be interpreted to mean that the relative opportunity costs of some of the cells are wrong.* To correct this situation we must obtain a set of three independent zero cells by revising the current total-opportunity-cost matrix. The result of the revision, therefore, must be such that either one of the zeros of the current total-opportunity-cost matrix (Table 10.11) is transferred to one of the uncovered nonzero cells or a new zero appears in one of the uncovered nonzero cells. We would, intuitively, want this cell to emerge as a new zero cell. The procedure for accomplishing this† consists in (1) subtracting the

---

* We know that the assignment problem involves decision making under certainty, and hence there must be at least one strategy (assignment) that involves a total opportunity cost of zero.

† The rationale for this procedure is embedded in the mathematical theorems that will be presented in Section 10.11.

lowest entry in the uncovered cells of the total-opportunity-cost matrix from all the uncovered cells and (2) adding the same lowest entry to *only* those cells in which the covering lines of step 2 cross.

In our example, the lowest entry in the uncovered cells of the total-opportunity-cost matrix (Table 10.11) is 1 (cell $O_1 D_2$). Subtracting this from all the uncovered cells and adding it to only those cells (in this case, cell $(O_3 D_1)$) in which the covering lines of step 2 cross, we obtain a revised total-opportunity-cost matrix (see Table 10.12b). An application of the test of step 2 to the revised total-opportunity-cost matrix shows that the minimum number of lines needed to cover all the zeros is 3.* Since the number of rows of this matrix is also 3, an optimal assignment can be made.

Table 10.12a

	D_1	D_2	D_3	
O_1	0	1	8	
O_2	0	2	4	
O_3	4	0	0	→ covering line 1

covering line 2

Table 10.12b

	D_1	D_2	D_3
O_1	0	0	7
O_2	0	1	3
O_3	5	0	0

Once it is established that an optimal assignment can be made, we search for a row or column in which there is only one zero cell. The first assignment is made to that zero cell, and the row and column in which this cell lies are crossed out. The remaining rows and columns of the matrix are examined, to find that row or column in which there remains only one zero cell. Another assignment is made, and the respective row and column are crossed out. The procedure is repeated until a complete assignment has been made. The optimal-assignment sequence for our problem is shown in Table 10.13. Since cell $O_3 D_3$ is the only zero cell in column D_3, we make the first assignment to cell $O_3 D_3$ and cross out row O_3 and column D_3.† In the

Table 10.13a

	D_1	D_2	D_3
O_1	0	0	7
O_2	0	1	3
O_3	5	0	×0

Table 10.13b

	D_1	D_2	D_3
O_1	0	×0	7
O_2	0	1	3
O_3	5	0	0

Table 10.13c

	D_1	D_2	D_3
O_1	0	0	7
O_2	×0	1	3
O_3	5	0	0

* Each line must be drawn in such a manner that it covers the largest number of zeros in the matrix.

† The making of an assignment is shown by placing the symbol × in the appropriate cell.

reduced matrix, we note that cell O_1D_2 is the only zero cell in column D_2. Hence we make the second assignment to cell O_1D_2 and cross out row O_1 and column D_2. This leaves only one zero cell open (cell O_2D_1), and therefore the third assignment is made to that cell. Thus we have the following optimal assignment:

Assign job O_1 to machine D_2.
Assign job O_2 to machine D_1.
Assign job O_3 to machine D_3.

The total opportunity cost associated with this optimal assignment is, of course, zero. The total cost of this assignment, as can be easily verified from the original cost matrix, is $49.

10.8   PROCEDURE SUMMARY FOR THE ASSIGNMENT METHOD (MINIMIZATION CASE)

STEP 1. Determine the total-opportunity-cost matrix.

a. Arrive at a column-opportunity-cost matrix by subtracting the lowest entry of each column of the given payoff matrix from all the entries in its column.

b. Then subtract the lowest entry of each row of the matrix obtained in step 1a from all the entries in its row.

The result of step 1b gives the total-opportunity-cost matrix.

STEP 2. Test for optimality (determine whether an optimal assignment can be made).

a. Cover all the zeros of the current total-opportunity-cost matrix with the *minimum* possible number of horizontal and vertical lines.

b. If the number of lines drawn in step 2a equals the number of rows (or columns) of the matrix, the problem can be solved. Make a complete assignment so that the total opportunity cost involved in the assignment is zero.

c. If the number of lines drawn in step 2a is less than the number of rows (or columns) of the matrix, proceed to step 3.

STEP 3. Revise the total-opportunity-cost matrix.

a. Subtract the lowest entry in the uncovered cells of the current total-opportunity-cost matrix from all the uncovered cells.

b. Add the same lowest entry to *only* those cells in which the covering lines of step 2 cross.

The result of steps 3a and 3b is a revised total-opportunity-cost matrix.

STEP 4

Repeat steps 2 and 3 until an optimal assignment, which has a total opportunity cost of zero, can be made.

10.9 THE ASSIGNMENT METHOD (MAXIMIZATION CASE)

Except for one transformation, an assignment problem in which the objective is to maximize the total payoff can be solved by the assignment algorithm presented in Section 10.8. The transformation involves subtracting all the entries of the original payoff matrix from the highest entry of the original payoff matrix. The transformed entries give us the "relative costs," and the problem then becomes a minimization problem. Once the optimal assignment for this transformed minimization problem has been identified, the total value of the optimal assignment can be found by adding the original payoffs for those cells to which the assignments have been made.

10.10 MULTIPLE OPTIMAL SOLUTIONS

After having established the fact that an optimal solution for an assignment problem exists, we may find that the number and positions of the zero cells in the final total-opportunity-cost matrix are such that more than one optimal assignment can be made. The presence of multiple optimal solutions for an assignment problem can be identified by the fact that, while assignments are being made via the final* total-opportunity-cost matrix, a row or column that contains only one zero cell cannot be located. Let us illustrate by considering the assignment problem in Table 10.14.

Table 10.14

	D_1	D_2	D_3	D_4	D_5
O_1	2	4	3	5	4
O_2	7	4	6	8	4
O_3	2	9	8	10	4
O_4	8	6	12	7	4
O_5	2	8	5	8	8

By subtracting the lowest entry in each column from all the entries in its column (step 1a), we obtain the column-opportunity-cost matrix (see Table 10.15). In this particular problem, the column-opportunity-cost matrix is also the total-opportunity-cost matrix. When we subtracted the lowest entry in each column from all the entries in its column, we obtained a

---

* Such as the revised total-opportunity-cost matrix of Table 10.12b. It is final in the sense that we know that this matrix can guide us in making an optimal assignment, and thus it need not be revised.

Table 10.15

	D_1	D_2	D_3	D_4	D_5
O_1	0	0	0	0	0
O_2	5	0	3	3	0
O_3	0	5	5	5	0
O_4	6	2	9	2	0
O_5	0	4	2	3	4

zero cell in each row of the matrix. Thus, if we now subtract the lowest entry in each row from all the entries in its row (step 1b), the matrix will not be changed. Hence the matrix of Table 10.15 is indeed the total-opportunity-cost matrix.

Next, we must draw the minimum number of horizontal and vertical lines needed to cover all the zeros of the total-opportunity-cost matrix (step 2). The particular sequence in which these lines must be drawn in our example is shown in Table 10.16.* Since we need only four lines to cover all the zeros of the total-opportunity-cost matrix, whereas the number of rows in the matrix is 5, an optimal assignment cannot be made at this stage.

Table 10.16

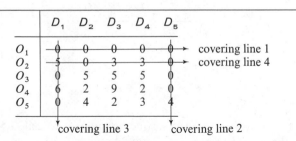

The next step is to revise this total-opportunity-cost matrix by the application of step 3 of the assignment algorithm. An examination of the uncovered cells of the current total-opportunity-cost matrix (Table 10.16) indicates that the smallest entry in the uncovered cells is 2. We, therefore, subtract 2 from all the uncovered cells and add 2 to only those cells in which the covering lines cross (O_1D_1, O_1D_5, O_2D_1, and O_2D_5). Thus the revised total-opportunity-cost matrix in Table 10.17 is obtained. An attempt to draw lines to cover the zeros of the matrix shows that a minimum number of 5 lines is required. The number of rows in the matrix is also 5; so we conclude that an optimal assignment can now be made.

* Line 4 could have been drawn to cover column D_2 rather than row O_2. This, however, would have made no difference in the result of this step.

Table 10.17

	D_1	D_2	D_3	D_4	D_5
O_1	2	0	0	0	2
O_2	7	0	3	3	2
O_3	0	3	3	3	0
O_4	6	0	7	0	0
O_5	0	2	0	1	4

An examination of the rows and columns of the matrix in Table 10.17 shows that only one row (row O_2) contains a single zero cell. Hence the first assignment is made to this cell ($O_2 D_2$). This means that row O_2 and column D_2 can be deleted from the matrix, and we are left with the matrix of Table 10.18. An examination of the reduced matrix shows that there is no row or column in which there is only one zero cell. This fact, as we mentioned earlier, indicates the presence of multiple optimal solutions in an assignment problem.

Table 10.18

	D_1	D_3	D_4	D_5
O_1	2	0	0	2
O_3	0	3	3	0
O_4	6	7	0	0
O_5	0	0	1	4

Let us arbitrarily make the second assignment to cell $O_1 D_3$. Then the remaining three assignments may be made in this order: third assignment to cell $O_5 D_1$, fourth assignment to cell $O_4 D_4$, and fifth assignment to cell $O_3 D_5$. On the other hand, if the second assignment had been made to, say, cell $O_3 D_1$, the remaining three assignments would have been to cells $O_5 D_3$, $O_4 D_5$, and $O_1 D_4$. The total costs of these alternative optimal assignments are, of course, the same ($20). The two alternative optimal assignments are

First optimal assignment:

Assign O_2 to D_2.
Assign O_1 to D_3.
Assign O_5 to D_1.
Assign O_4 to D_4.
Assign O_3 to D_5.

Second optimal assignment:

Assign O_2 to D_2.
Assign O_3 to D_1.
Assign O_5 to D_3.
Assign O_4 to D_5.
Assign O_1 to D_4.

A little reflection will show that, for the assignment problem under consideration, there are many more alternative optimal solutions.

10.11 MATHEMATICAL FOUNDATION OF THE ASSIGNMENT ALGORITHM

The foundation of the assignment algorithm is provided by (1) a mathematical theorem proved by the Hungarian mathematician Konig, and (2) a statement of a certain matrix property.*

The theorem states: *If the elements of a matrix are divided into two classes by a property R, then the minimum number of lines that contain all the elements with the property R is equal to the maximum number of elements with the property R, with no two on the same line.*

The particular matrix property is: *Given a cost matrix* $A = [a_{ij}]$, *if we form another matrix* $B = [b_{ij}]$, *where* $b_{ij} = a_{ij} - u_i - v_j$, *and where* u_i *and* v_j *are arbitrary constants, the solution of* A *is identical with that of* B.

To clarify the relationship of the solution stages based on the above-mentioned theorem and matrix property to the corresponding steps of the assignment algorithm, we shall solve the problem of Table 10.4 by these two approaches.

Solution by the Assignment Algorithm

Original Cost Matrix

	D_1	D_2	D_3
O_1	20	27	30
O_2	10	18	16
O_3	14	16	12

Solution Based on the Konig Theorem and the Matrix Property

Original Cost Matrix

		D_1	D_2	D_3
	O_1	20	27	30
$A = [a_{ij}] =$	O_2	10	18	16
	O_3	14	16	12

* The mathematical theorem, the matrix property, and references are given in Churchman, et al. [1957, Chapter 12].

STEP 1

a. Subtract the lowest entry in each column of the original cost matrix from all the entries in its column. The above matrix becomes

	D_1	D_2	D_3
O_1	10	11	18
O_2	0	2	4
O_3	4	0	0

b. Subtract the lowest entry in each row of the matrix obtained in step 1a from all the entries in its row. The above matrix becomes

	D_1	D_2	D_3
O_1	0	1	8
O_2	0	2	4
O_3	4	0	0

STEP 1

a. Transform the given payoff matrix by the relationship $b_{ij} = a_{ij} - u_i - v_i$. In particular, let $u_1^{(0)} = u_2^{(0)} = u_3^{(0)} = 0$, and let $v_1^{(0)} = 10, v_2^{(0)} = 16,$ and $v_3^{(0)} = 12.$* Then

	D_1	D_2	D_3	$u_i^{(0)}$
O_1	20	27	30	0
O_2	10	18	16	0
O_3	14	16	12	0
$v_j^{(0)}$	10	16	12	

$$\Rightarrow [b_{ij}] = \begin{array}{|c|c|c|} \hline 10 & 11 & 18 \\ \hline 0 & 2 & 4 \\ \hline 4 & 0 & 0 \\ \hline \end{array}$$

b. Transform the matrix derived by the relationship $b'_{ij} = b_{ij} - u_i - v_j$. In particular, let $u_1^{(1)} = 10, u_2^{(1)} = 0, u_3^{(1)} = 0,$ and $v_1^{(1)} = v_2^{(1)} = v_3^{(1)} = 0.$

	D_1	D_2	D_3	$u_i^{(1)}$
O_1	10	11	18	10
O_2	0	2	4	0
O_3	4	0	0	0
$v_j^{(1)}$	0	0	0	

$$\Rightarrow [b'_{ij}] = \begin{array}{|c|c|c|} \hline 0 & 1 & 8 \\ \hline 0 & 2 & 4 \\ \hline 4 & 0 & 0 \\ \hline \end{array}$$

* u_i's refer to the row numbers, and v_j's refer to the column numbers. Furthermore, the subscripts refer to the particular row or column, whereas the superscripts refer to the sequence of transformations. For example, $u_1^{(0)}$ is the value to be subtracted from all the entries of the first row of the original matrix. Similarly, $v_1^{(1)}$ would be the value to be subtracted from all the entries of the first column of the matrix derived after the first iteration, and so on.

STEP 2

Cover all the zeros of the total-opportunity-cost matrix with the minimum possible number of horizontal and vertical lines.

Since only two lines are needed to cover all the zeros, whereas the number of rows of the matrix is 3, the optimal assignment cannot be made at this stage

Proceed to step 3.

STEP 2

An application of the Konig theorem to this matrix $[b'_{ij}]$ shows that, since the minimum number of lines to cover the zero cells is 2, the maximum number of real zero cells is also 2. However, we need three real zero cells to make an optimal assignment. Hence the matrix must be further transformed. Note that property R in the theorem corresponds to a total opportunity cost of zero in a given cell.

STEP 3

Subtract the lowest entry in the uncovered cells from all the uncovered cells, and add it to only those cells in which the covering lines of step 2 cross. The matrix of step 1 becomes

	D_1	D_2	D_3
O_1	0	0	7
O_2	0	1	3
O_3	5	0	0

STEP 3

Transform the previous matrix $\mathbf{B'} = [b'_{ij}]$ by the relationship $b''_{ij} = b'_{ij} - u_i - v_j$. In particular, let $u_1^{(2)} = 1$, $u_2^{(2)} = 1$, $u_3^{(2)} = 0$, and let $v_1^{(2)} = -1$, $v_2^{(2)} = 0$, and $v_3^{(2)} = 0$. Then

	D_1	D_2	D_3	$u_i^{(2)}$
O_1	0	1	8	1
O_2	0	2	4	1
O_3	4	0	0	0
$v_j^{(2)}$	-1	0	0	

$$\Rightarrow [b''_{ij}] = \begin{array}{|c|c|c|} \hline 0 & 0 & 7 \\ \hline 0 & 1 & 3 \\ \hline 5 & 0 & 0 \\ \hline \end{array}$$

STEP 4

Since it takes a minimum of three lines to cover all the zero cells of the above total-opportunity-cost matrix and the number of rows is also 3, an optimal assignment that

STEP 4

An application of the Konig theorem to this matrix $[b''_{ij}]$ shows that, since the minimum number of lines needed to cover the zero cells is 3, the maximum number of real

has a total opportunity cost of zero can be made. The optimal assignment is O_2 to D_1, O_1 to D_2, and O_3 to D_3.

zero cells is also 3. Hence an optimal assignment can be made. The optimal assignment is O_2 to D_1, O_1 to D_2, and O_3 to D_3.

Step-by-step comparison of the two approaches for solving an assignment problem shows that the assignment algorithm, in essence, repeatedly applies the Konig theorem to the given payoff matrix until a set of independent zeros is discovered. It is hoped that the comparison of the two approaches will give the reader a thorough understanding of the assignment algorithm.

It may be of interest to point out that the u_i's and v_j's of the mathematical theorem are the dual variables corresponding to the primal assignment problem. They also correspond to the row and column numbers of the transportation method.

REFERENCES

See references at the end of Chapter 7.

REVIEW QUESTIONS AND PROBLEMS

10.1. In what sense is the assignment model a special case of the transportation model and of the general linear programming model?

10.2. The assignment problem is inherently degenerate. Explain.

10.3. Describe the approach of the assignment model.

10.4. Under what conditions can the assignment problem have multiple optimal solutions?

10.5. An engineering manager has four process engineers to prepare four job proposals. The estimated number of hours that each engineer will spend on different proposals is shown in Table 10.19.

(a) If each engineer costs the company $10 per hour, find the optimal assignment.

(b) What will be the total cost?

Table 10.19

	Proposal			
Engineer	1	2	3	4
A	30	25	60	42
B	35	30	50	37
C	32	28	55	45
D	40	20	60	42

10.6. Five jobs are to be processed on five different machines. Any job can be completely processed on any machine. The data on profit obtained from processing each job on a specific machine are shown in Table 10.20.

Table 10.20

Job	Machine				
	1	2	3	4	5
A	76	74	84	76	86
B	80	72	80	66	68
C	72	60	66	50	54
D	74	56	56	48	70
E	80	42	72	74	80

(a) Which job should be processed on which machine?
(b) What is the maximum profit?

10.7. Four teachers are to be assigned to teach four courses on a one-to-one basis. The effectiveness of each teacher to teach a specific course, based on his or her qualifications and talents, is estimated as shown in Table 10.21 (effectiveness is measured on a ratio scale).

Table 10.21

Teacher	Course			
	1	2	3	4
A	200	115	190	175
B	180	140	110	130
C	210	140	90	120
D	170	160	150	100

(a) Find the optimal assignment.
(b) Can such a model be used in practice?

10.8. A company has just interviewed four salesmen for four jobs. Because of the differences in education and experience of the salesmen, it is anticipated that each salesman will not be equally effective in each sales territory. The estimated annual sales in thousands of dollars for each salesman–territory assignment are shown in Table 10.22. Determine which salesman the company should hire and how they (salesmen) should be assigned.

Table 10.22

	Sales Territory			
Salesman	1	2	3	4
A	30	27	31	39
B	28	18	28	37
C	33	17	29	41
D	27	18	30	43

10.9. The Flying Colors Airline owns four aircraft and has flights 1, 2, 3, and 4 from New York to Houston, and flights A, B, C, and D from Houston to New York. By assigning an aircraft to any pairing of flights 1, 2, 3, or 4 with flights A, B, C, or D, the airline can calculate the nonflying hours for the aircraft for that particular pairing. Table 10.23 gives the nonflying hours for each pairing.

Table 10.23

Houston to New York	New York to Houston			
	1	2	3	4
A	16	14	15	18
B	12	13	16	14
C	14	13	11	12
D	16	18	15	17

(a) Find the pairings of flights between New York and Houston that will optimize (minimize) the total nonflying hours of all four aircraft.
(b) What is the second-best pairing?

11
Integer Programming

11.1 INTRODUCTION

As discussed in Chapter 2, linear programming is only one component of the broad field known as mathematical programming. Nonlinear programming, stochastic programming, integer programming, and dynamic programming are some examples of programming methods other than linear programming. Each programming method, or model, represents an efficient optimization technique to solve a problem with a specific structure. And the specific structure in each category is determined by a set of assumptions. For example, the structure of the general linear programming model (Chapters 1 and 2) is based essentially on the following five important assumptions:

1. *Linearity* (the objective function as well as all the constraints are assumed to be linear).
2. *Certainty* (we assume complete knowledge regarding strategies, resources, technology, and the consequences or payoffs associated with each strategy).
3. *Divisibility* (all variables are assumed to be continuous and hence they can assume integer as well as fractional values).
4. *Nonnegativity* (all variables must assume a nonnegative value; that is, an absolute lower bound is specified on all variables).
5. *Single stage* (the problem assumes a fixed planning horizon consisting of only one stage).

What will happen if we make the general linear programming model *more* or *less* restrictive? That is, what kind of a modified model will evolve, and how shall we solve the modified model when we make some additional demands on, or assumptions in, the general linear programming model to more accurately reflect the real-life problems? What, for example, will emerge if we impose on the general linear programming model the restriction that some or all variables can only assume nonnegative integer values.* As discussed in Section 11.3, the additional constraint that requires some or all variables to assume only nonnegative integer values gives rise to the integer (linear) programming problem.

The purpose of this chapter is to describe integer programming and illustrate the bound-and-branch method of solving an integer programming problem. In Section 11.2 we shall examine some of the assumptions of linear programming and describe how, by modifying these assumptions, we enter the domain of some important mathematical programming methods. In Section 11.3 we explain the idea of rounding off a noninteger solution, present a graphical analysis of a simple integer programming problem, and illustrate the branch-and-bound method.

11.2 EXTENSIONS OF LINEAR PROGRAMMING

By *extensions* of linear programming we shall mean any model that emerges when one (or more) of the assumptions of the general linear programming model is modified to make the model either more or less restrictive. Two examples of such additional (or modifying) assumptions were discussed in Chapter 9 (the transportation model) and Chapter 10 (the assignment model). The transportation model was developed by imposing on the general linear programming model the two additional requirements: (1) we are dealing with a homogeneous commodity, and (2) each input–output co-efficient (a_{ij}) is either 1 or 0. The assignment model, as the reader will recall, introduced the additional restriction that each variable (x_{ij}) must take a value of either 1 or 0. As discussed in Chapters 9 and 10, these additional restrictions, although they could be directly met by the general linear programming model, were more efficiently handled by *special-structure* linear programming models.

Let us now examine a few modifications in the assumptions of the general linear programming model that lead to some of the most important mathematical programming models. We examine the four assumptions (linearity, certainty, divisibility, and nonnegativity) and make some observations.

If the assumption of linearity were to be modified by assuming the existence of a nonlinear objective function or nonlinear constraints, we enter

---

* An integer is a whole number as distinguished from a fraction. The numbers 0, 1, 2, . . . are nonnegative integers.

the domain of *nonlinear programming*. A nonlinear programming problem arises whenever we have either a nonlinear objective function, or a set of nonlinear constraints, or both.*

If the assumption of certainty is modified by assuming that some or all parameters of the problem are described by random variables (that is, they are probabilistic rather than deterministic), we are in the domain of *stochastic programming*. The idea of stochastic programming is to convert the probabilistic problem into an equivalent deterministic problem, and then solve the problem.†

If the assumption of divisibility is modified by restricting some or all variables to integer values, we are in the domain of *integer programming*. In Section 11.3 we discuss integer programming and illustrate the branch-and-bound method of solving an integer programming problem.‡

The simple nonnegativity assumption imposes no upper bound on the values of the variables. However, in many real-life problems it may be necessary to specify the *upper bounds* for some or all variables. When this happens, the linear programming problem is transformed into what is known as the *bounded-variable problem*. An efficient algorithm, which is based on the simplex method, has been developed to solve the bounded-variable problem.** It should also be observed that the general linear programming model can handle unrestricted variables X_j by a simple transformation $X_j = Y_j - W_j$; with $Y_j, W_j \geq 0$.

If we go beyond the single-stage assumption of linear programming and consider those problems which require not one but a *sequence* of decisions, we are in the domain of *dynamic programming*. In dynamic programming the planning horizon is divided into more than one stage (*multistage* as opposed to single-stage), and the outcome of a decision at one stage affects the subsequent decision for the next stage, and so on. There are a host of business and economic problems that are multistage (or multiperiod), and they can be solved by dynamic programming.§

11.3 INTEGER PROGRAMMING

Linear programming models assume divisibility. That is, all the variables are assumed to be continuous, and hence they can be assigned any nonnegative integer as well as fractional value. However, in many problems the

* For a brief discussion of nonlinear programming, see Chapter 7 of Loomba and Turban [1974].

† For a brief and simple illustration, see Wagner [1970, pp. 351–59], and Taha [1971, pp. 649–53].

‡ For some other methods of solving integer programming problems, see Chapter 6 of Loomba and Turban [1974].

** For a simple illustration, see Chapter 10 of Chung [1963].

§ See Chapter 8 of Loomba and Turban [1974]. The chapter (89 pages long) illustrates dynamic programming by employing a variety of illustrative examples.

assumption of divisibility does not reflect the realities of life. For example, we cannot build 2.2 plants or 2.25 machines. Nor does an optimal solution that asks for the scheduling of 3.5 flights has any operational meaning. We must schedule either 3 or 4 flights. Similarly, we can cite several other problems or examples in which the decision variables must assume only *integer* values. Thus there is the need to impose, on the general linear programming model, an additional constraint that some or all of the variables can assume only integer values. When this is done, the resulting model is termed the integer (linear) programming model. It consists of four components: (1) a linear objective function, (2) a set of linear constraints, (3) a set of nonnegativity constraints, and (4) integer-value constraints for some or all variables. When all variables are restricted to integer values, we have an *all-integer problem*. If only some of the variables are restricted to integer values, we have a *mixed-integer problem*.

Various methods have been developed to solve integer programming problems.* As compared to the simplex algorithm, the computational effort required for solving integer programming problems is rather extensive. For this reason, we shall restrict our illustration to only one of the algorithms, the branch-and-bound method. However, we first make a few comments on the idea of rounding off a noninteger solution and present a graphical analysis of a simple integer programming problem.

11.3.1   Rounding Off the Noninteger Solution

One obvious approach for obtaining an integer solution is to first arrive at the optimal noninteger solution and then round off the values of the noninteger basic variables to their nearest value. The major advantage of this approach is the economy of time and cost, since the alternative of solving the problem by a regular integer programming algorithm will often require additional computations. The major disadvantages of this approach are that (1) the rounded solution might not be the real optimal integer solution, and (2) the rounded solution might be an infeasible solution. Hence, when using the rounding-off approach, *each* rounded solution must be carefully tested for feasibility.

Let us consider the optimal noninteger solution to our familiar product-mix problem shown in Table 3.2 and stated in (11.2). The noninteger optimal solution is

$$X = 185\tfrac{5}{7}$$
$$Y = 107\tfrac{1}{7}$$
$$\text{profit} = \$7,700$$

---

* These seven are discussed in Loomba and Turban [1974]: (1) rounding off a noninteger solution, (2) complete enumeration, (3) graphical approach, (4) Gomory's method, (5) Land and Doig's method, (6) branch-and-bound method, and (7) heuristic programming.

We shall now examine four rounding alternatives, carefully testing each for feasibility (see Table 11.1). Note that the $7,688 profit associated with the optimal integer solution is less than the $7,700 profit associated with the optimal noninteger solution. The difference (in this case, $12) between the objective function of the optimal noninteger solution and the optimal integer solution represents the *cost of indivisibility*.

Table 11.1

	Rounding Alternative	Is the Rounded Solution Feasible?	Profit	
Case I:	$X = 186$ $Y = 107$	No		
Case II:	$X = 185$ $Y = 108$	No		
Case III:	$X = 185$ $Y = 107$	Yes	$7,679	
Case IV:	$X = 184$ $Y = 108$	Yes	$7,688 ◄—	optimal integer solution

11.3.2 Graphical Analysis

Let us analyze the following integer programming problem by utilizing its graphical representation.

Maximize $\qquad 6X + 4Y$

subject to the constraints

$$
\begin{aligned}
1.2X + 2Y &\leqslant 14.4 \\
X \quad\ \ &\leqslant 8 \qquad\qquad (11.1)\\
Y &\leqslant 4
\end{aligned}
$$

and

$$X, \quad Y \geqslant 0, \text{ and integer}$$

The integer programming problem given by (11.1) represents a managerial decision in which an optimal *integer* mix of two types of equipment is to be determined, assuming that the payoffs (objective function coefficients) associated with each unit of the two types of equipment are known.

Let us start by finding the optimal solution to the problem stated in (11.1) by the graphical method presented in Chapter 3. Shown in Figure 11.1 are the

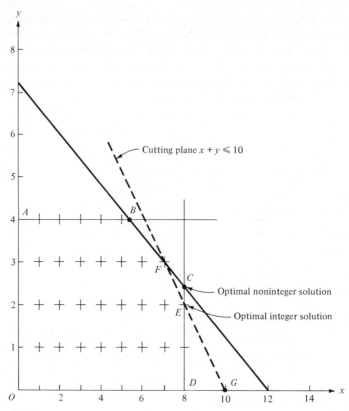

Figure 11.1

relevant constraints, the feasible convex set $OABCD$, and the optimal solution point C. The optimal *noninteger* solution is

$$X = \quad 8$$
$$Y = \quad 2.4$$
$$\text{profit} = \$57.60$$

The optimal integer solution to this problem, as the reader can verify, is $X = 8$, $Y = 2$, (point E) and profit $= \$56$. Let us examine Figure 11.1 in order to get a "feel" of the integer programming problem and see how some of the integer programming algorithms operate.

First, it should be noted that while *any* point in the convex set $OABCD$ can be a feasible linear programming solution, only the points shown by the lattice points can be feasible integer solutions.* Second, if the simplex algorithm is to be used in integer programming, the convex set $OABCD$

---

* *Lattice points*, marked by the sign $+$ in Figure 11.1, are those whose coordinates are integer numbers. Whole numbers on either of the axes are obviously integer.

should, somehow, be reduced in such a manner that the *extreme* (or corner) *points* of the reduced convex set become all-integer points. This reduction, for example, can be accomplished by constructing one or more "*cutting planes.*" We have drawn a cutting plane *GEF* in Figure 11.1 that represents the constraint $X + Y \leqslant 10$.* These cutting planes are generated by an integer programming algorithm known as *Gomory's method*.† The heart of Gomory's method is to construct a convex set (with a minimum area) that covers all the integer values (lattice points) of the original noninteger convex set. This is accomplished by constructing the cutting planes that are imposed, *one at a time*, on the original convex set. The cutting planes reflect the additional linear constraints (called *Gomorian constraints*) that are added to the problem in order to converge on the integer solution.

The graphical analysis points up the important fact that the set of feasible *integer* solutions is always less than or equal to the set of feasible solutions without the integer constraints. This means that the optimal value of an integer solution will always be less than or equal to the optimal value of a noninteger solution to the same problem. Hence, as shown in Section 11.3.1, the manager must pay a price for imposing the indivisibility requirement.

11.3.3 The Branch-and-Bound Method

The *branch-and-bound algorithm* starts by solving the given linear programming problem without the integer constraints. If the resulting optimal solution satisfies the integer requirements, the problem is solved. Otherwise, the original problem is split into two *branches* by aiming the search at two integer values that are immediately below and above some *noninteger basic variable* that needs to be integerized.

If, after the first branching, an optimal integer solution is obtained, the search is stopped. Otherwise, further branching takes place, creating more nodes‡ and more descendant problems. The process of branching creates, at *each* node, two descendant problems. The descendant problems, emanating from each node, are again solved. The search is finally stopped when, from among the optimal solutions, the integer solution with the highest objective function (maximization case) has been identified.

Assume that the optimal noninteger value of a basic variable Y is 2.4. Then, one descendant problem will contain the additional constraint $Y \leqslant 2$, and the second descendant problem will contain the additional constraint $Y \geqslant 3$. As, and if, further branching is needed, the new constraint (say $Y \leqslant 2$) must be included in *all subsequent descendants* of that branch.

---

* This constraint ($X + Y \leqslant 10$), which represents the cutting plane, was derived by using Gomory's method. In the simple problem shown in Figure 11.1, it is not difficult to argue that a cutting plane drawn through the lattice points E and F would indeed be the minimum convex set that covers all the integer solutions.

† Gomory's method is illustrated in Section 6.5 of Loomba and Turban [1974].

‡ A *node* is the point where the problem is split into two branches.

Similarly, as and if needed, the other new constraint ($Y \geqslant 3$) must be included in *all subsequent descendants* of the second branch.

We have now explained the "branch" part of the branch-and-bound method. The term *bound* refers to the fact that the optimal integer solution to a maximizing problem is always less than or equal to the optimal non-integer solution (thus there exists an upper bound). Conversely, the optimal integer solution to a minimizing problem is always greater than or equal to the optimal noninteger solution (thus there exists a lower bound). The existence of these upper and lower bounds is used to decide whether it is necessary to further branch from a given branching node.

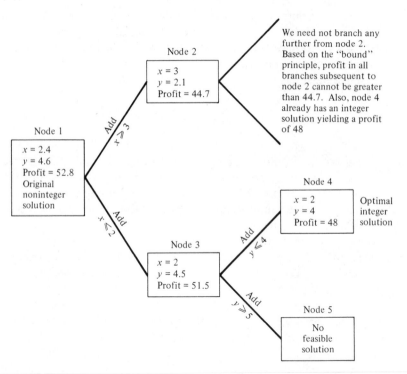

Figure 11.2

Consider the hypothetical case shown in Figure 11.2. As stated in Figure 11.2, no further branching from node 2 is needed. This is because any descendant problems from node 2 will have an upper bound of 44.7. And we already have an integer solution at node 4 that yields a profit of \$48.

We now proceed to illustrate the branch-and-bound method by solving the problem stated in (11.2).

Maximize $\qquad\qquad 23X + 32Y$

subject to the constraints

$$\begin{aligned} 10X + 6Y &\leqslant 2{,}500 \\ 5X + 10Y &\leqslant 2{,}000 \\ X + Y &\leqslant 500 \end{aligned}\qquad(11.2)$$

and

$$X, \qquad Y \geqslant \qquad 0,\ \text{and integer}$$

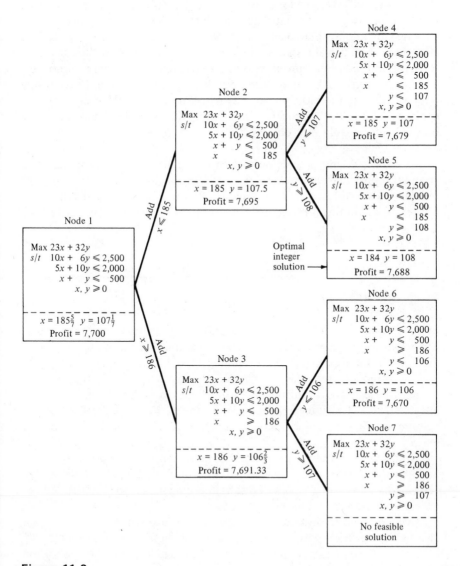

Figure 11.3

The optimal noninteger solution, shown at node 1 in Figure 11.3, is

$$X = 185\tfrac{5}{7}$$

$$Y = 107\tfrac{1}{7}$$

$$\text{profit} = 7,700$$

Since the optimal solution is noninteger, we must branch by first selecting
either X or Y to be integerized. Let us select the noninteger basic variable
X for branching. We now form two descendant problems, shown at nodes 2
and 3, respectively. Note that the descendant problem stated at node 2 was
obtained by adding the constraint $X \leqslant 185$ to the original problem. The
descendant problem shown at node 3 was obtained by adding the constraint
$X \geqslant 186$ to the original problem. The two descendant problems at nodes 2
and 3 are solved and, as can be seen from their optimal solutions (listed below
the dashed line), the basic variable Y is still noninteger. Hence we branch
again by adding the appropriate constraints as shown in Figure 11.3. The
optimal integer solution, obtained at node 5, is $X = 184$, $Y = 108$, and
profit $= \$7,688$.

The branch-and-bound method can be efficiently programmed on a
computer, provided that the integer requirement does not apply to a large
number of variables. A brief discussion of the computational aspects of
integer programming is given in Section 6.13 of Loomba and Turban [1974].

11.4 SUMMARY

In this chapter we have very briefly described the nature of emerging models
when some of the assumptions of linear programming are modified. One of
these models is the integer programming model. The integer (linear) programming model arises when we impose, on the general linear programming
model, the restriction that some or all of the variables can only assume
integer values. The integer programming model accurately reflects many
real-life problems in which fractional answers do not make any operational
sense.

Integer programming problems can be solved by a variety of methods
or algorithms. In this chapter we examined the rounding-off approach,
described a graphical analysis of a simple problem, and then illustrated the
widely used branch-and-bound method.

Managers are concerned not only with the building of models but
also with the important question of application and implementation.
In Chapter 12 we shall examine some of the issues involved in the application
and implementation aspects of management science models.

REFERENCES

Baumol, W. J. *Economic Theory and Operations Analysis*, 2nd ed. Englewood Cliffs, N.J.: Prentice-Hall, Inc., 1965.

Beale, E. M. L. "Survey of Integer Programming." *Operational Research Quarterly*, **16**, (June 1965), 219–28.

Beale, E. M. L. *Applications of Mathematical Programming Techniques*. New York: American Elsevier Publishing Co., Inc., 1970.

Childress, R. L. *Sets, Matrices, and Linear Programming*. Englewood Cliffs, N.J.: Prentice-Hall, Inc., 1974.

Chung, An-min. *Linear Programming*. Columbus, Ohio: Merrill, Inc., 1963.

Cooper, L., and D. Steinberg. *Linear Programming*. Philadelphia: W. B. Saunders Company, 1974.

Daellenbach, H. G., and E. G. Bell. *User's Guide to Linear Programming*. Englewood Cliffs, N.J.: Prentice-Hall, Inc., 1970.

Gomory, R. E. "An Algorithm for Integer Solutions to Linear Programs." In Robert L. Graves and Philip Wolfe (eds.), *Recent Advances in Mathematical Programming*. New York: McGraw-Hill Book Company, 1963.

Haley, K. B. *Mathematical Programming for Business and Industry*. New York: St. Martin's Press, Inc., 1967.

Hillier, F. S., and G. J. Lieberman. *Introduction to Operations Research*, 2nd ed. San Francisco: Holden-Day, Inc., 1974.

House, W. C. *Operations Research—An Introduction to Modern Applications*. Philadelphia: Auerbach Publishers, Inc., 1972.

Hughes, A. J., and D. Grawiog. *Linear Programming*. Reading, Mass.: Addison-Wesley Publishing Co., Inc., 1973.

Land, A. H., and A. G. Doig. "An Automatic Method of Solving Discrete Programming Problems." *Econometrica*, **28** (July 1960), 497–520.

Lawler, E. L., and D. E. Wood, "Branch-and-Bound Methods: A Survey." *Operations Research*, **14** (July–August 1966), 699–719.

Loomba, N. P., and E. Turban. *Applied Programming for Management*. New York: Holt, Rinehart and Winston, Inc., 1974.

McMillan, C. *Mathematical Programming*. New York: John Wiley & Sons, Inc., 1974.

Simmons, D. M. *Linear Programming for Operations Research*. San Francisco: Holden-Day, Inc., 1972.

Spivey, W. A., and R. M. Thrall. *Linear Optimization*. New York: Holt, Rinehart and Winston, Inc., 1970.

Strum, J. E. *Introduction to Linear Programming*. San Francisco: Holden-Day, Inc., 1972.

Taha, H. A. *Operations Research: An Introduction*. New York: Macmillan Publishing Co., Inc., 1971.

Wagner, H. M. *Principles of Management Science with Applications to Executive Decisions*, 2nd ed. Englewood Cliffs, N.J.: Prentice-Hall, Inc., 1975.

Wagner, H. M. *Principles of Operations Research with Applications to Managerial Decisions*, 2nd ed. Englewood Cliffs, N.J.: Prentice-Hall, Inc., 1975.

REVIEW QUESTIONS AND PROBLEMS

11.1. Explain, in your own words, how the convex set of noninteger solutions can be reduced to a convex set whose corner points are all integer points.

11.2. What are some of the models that emerge when further conditions are imposed on the five assumptions of the general linear programming model given in Section 11.1?

11.3. What is the rationale of the branch-and-bound method?

11.4. Use the branch-and-bound approach to find an optimal integer solution to the following problem:

Minimize $8X + 7Y$

subject to the constraints

$$2X + 3Y \geqslant 45$$
$$3X + Y \geqslant 37$$
$$X, \quad Y \geqslant 0, \text{ and integer}$$

11.5. The AB Company produces two products, A and B, which have to go through three processes. The products give a profit of \$12 and \$14, respectively. The time requirements are given in Table 11.2. What product mix will give a maximum profit? Assume that the products are not divisible.

Table 11.2

	Product		
Process	A	B	Capacity
Metalwork	22	33	2,400
Assembly	40	30	2,400
Packaging	24	32	1,800

11.6. Given:

Maximize $20X + 10Y + 14Z$

subject to the constraints

$$10X + 3Y + 6Z \leqslant 100$$
$$10X + 6Y + 8Z \leqslant 120$$
$$8X + 9Y + 4Z \leqslant 150$$
$$X, \quad Y, \quad Z \geqslant 0, \text{ and integer}$$

Find the optimal integer solution.

11.7. The Air Force wishes to provide radar coverage for a 5,000-mile-wide area. It can install two types of radar stations; type A has a radius of 500 miles, type B

has a radius of 250 miles. The costs and expenditures for each station are listed in Table 11.3.

Man-hours limitation: 2,700 hours
Total energy available: 4,500 megawatts

Table 11.3

Type	Cost	Man-Hours Required	Energy Required (megawatts)
A	$36,000,000	425	750
B	$17,500,000	300	400

Find an all-integer solution to this problem which minimizes the cost of providing coverage for the area.

11.8. An advertising agency plans to initiate a campaign. It intends to limit the campaign to newspaper ads and television broadcasts. Studies indicate that the two media will have the parameters (per advertisement) shown in Table 11.4. The campaign expenditures are limited to $45,000. It is also desired that (1) no more than 750,000 blue-collar workers be reached, (2) no less than 600,000 housewives be reached, (3) no less than 200,000 individuals in the suburbs be reached by either advertisement, and (4) no more than 550,000 individuals in the age group 16–21 be reached. Determine the optimal integer solution for the number and types of advertisements.

Table 11.4

	Newspaper	Television
Cost	$2,500	$7,500
No. of blue-collar workers	100,000	30,000
No. of housewives	50,000	275,000
No. in the suburbs	85,000	75,000
No. in the age group 16–21	35,000	200,000

11.9. A country wishes to perform an aerial survey of the entire nation. Two types of aircraft are available for the job. The first is a converted bomber that costs $485,000 per unit; the second is a converted light passenger plane that costs $93,000. The bomber can photograph 15,000 square miles per flight and can perform 2 flights per week. The second aircraft photographs 8,000 square miles per flight and can be expected to have four flights per week. The government has budgeted $10,000,000 for the purchase of the planes. Bombers may be purchased individually, but the passenger planes must be bought in groups of three. The area to be surveyed is 1,500,000 square miles. What is the best mix of aircraft?

11.10. A firm produces two electronic components, in lots of 1,000. The units (per 100) require the inputs shown in Table 11.5. Unit A contributes a profit of \$300 (per 100 units); unit B contributes a profit of \$175 (per 100 units). Determine the product mix in 1,000 units (integer solution).

Table 11.5

Component	Labor	Machine A	Machine B	Checking
A	10	12	40	4
B	15	9	27	7
Limits	450	425	900	275

12

Application and Implementation

12.1 INTRODUCTION

It is often said that the *process* is as important as the *product;* that *means* are
even more important than the *ends*; that *effort* is what really counts. These
statements, and the philosophy behind them, are important in any field of
human endeavor. They are, however, especially relevant to the task of
formulating management science models, deriving their solutions, and
implementing the programs, procedures, or policies suggested by the model
solutions. This is because management science models are invariably
quantitative models, and they require the problem to be expressed in quantitative and often mathematical terms. This requirement yields several dividends,
not the least of which is an increased understanding of the problem and a
more meaningful awareness of the system environment. Hence, whenever a
situation arises that calls for the application of linear programming, or any
other management science model, the first task is to define and formulate
the problem in quantitative terms. The very attempt to define the problem
in a precise and structured manner has several inherent advantages, as
described in Section 12.2. Yet, to derive full benefits from any model, tool, or
technique of modern management, the manager must move beyond the
stage of problem definition and formulation.* That is, the manager must
consider the very important aspect of *application* of the specific model and

---

* See Section 2.7 for a brief discussion of problem formulation.

implementation of the model solution. What, for example, is needed in terms of technical requirements and behavioral changes to transform a theoretical model into a practical reality? What technical support and facilities are required to prepare the problem data in the proper format, choose the proper computer program to execute the necessary computations, and arrange the computer output in a format that is meaningful in managerial decision making? What behavioral changes are desirable, and what organizational environment must be created to effectively implement the solution? These are obviously some of the most important questions and considerations that have real significance to the manager.

The technical problems related to what we have termed "application" have been given a great deal of attention in the management literature.* The behavioral problems concerning the implementation phase have only recently begun to draw the attention of management and behavioral scientists.† A detailed analysis of these problems is beyond the scope of this book. However, we shall present in this chapter a very brief discussion of those aspects of application and implementation phases that are most significant in determining the feasibility (in terms of technical requirements) and viability (in terms of behavioral requirements) of the success of linear programming models.‡

12.2 SOME IMPORTANT ASPECTS OF THE APPLICATION PHASE

One dimension of the application of linear programming is the capacity of the general linear programming model to solve various types of problems (e.g., scheduling, inventory, allocation) in such diverse sectors as business, government, education, military, and industry.** This dimension relates to the *scope* of applications. Another dimension concerns the various issues and aspects of application that arise in the case of a specific and individual situation. In this second dimension, the application aspects of linear programming can be considered at four different but related levels. At the *first* level is the requirement that the problem be defined and formulated in precise, quantitative terms, and that it be expressed mathematically. The completion of this requirement has several inherent advantages. For example, in the process of making a precise statement of the objective function, the manager will be forced to recognize, compare, and contrast the different objectives and goals of the system that is under analysis. At a very early stage of the analysis, the manager will have to decide whether he wishes to

---

* A brief historical development is given in Loomba and Turban [1974, p. 169]. Also, see Section 4.6 of Loomba and Turban [1974] and the references cited therein.

† See, for example, Ramsing and Moberg [1974], Hammond [1974], and Wagner [1971].

‡ Although the contents of this chapter were developed with reference to linear programming, they are obviously applicable to any management science model.

** For some interesting examples and exercises, see Ghosal et al. [1975].

formulate the problem in terms of a single objective, or must he simultaneously contend with two or more objectives and goals? If it is appropriate to formulate the problem in terms of a single goal, and if a clear identification of the various activities that influence the achievement of the goal can be made, the next task is to determine the *behavior* (is the objective function linear or nonlinear?) and the *relative influence* (i.e., the objective function coefficients) of the objective function. Thus the objective function could be linear or nonlinear, and the problem can call for the optimization of either a single goal or two or more goals. Linear programming deals with the optization of single goals, expressed in the form of linear objective functions. If the problem is such that multiple goals must be explicitly recognized, the manager must resort to some of the techniques discussed in Section 2.7.*

Similarly, the mere process of evolving a precise statement of the constraints will force the manager to think about improving technology, developing new contacts and sources for procuring organizational resources, developing new products, eliminating some of the current products, seeking new markets, and so on. The requirement and the exercise of problem formulation, therefore, yields several tangential but important benefits, increases the manager's understanding of his environment, sharpens his perception, and makes him a more effective manager. There is indeed a real, on-going, interaction among the requirement of making a mathematical statement of the problem, the selection of the appropriate model, and the effectiveness of managerial decision making.

At the *second* level of the application phase, we enter the technical world of the models and consider those problems where the linear programming model can be applied on an "as is" basis, without having to modify either the problem or the model. For example, if the manager recognizes a problem that corresponds exactly to one of the classical linear programming problems presented in Section 1.9 (e.g., the product-mix problem, the diet problem, the transportation problem, and the assignment problem), a straightforward application of the model can be contemplated.

At the *third* level are those situations where it is necessary, and possible, to either modify the problem so that it can then be solved by the linear programming model, or modify the linear programming model to solve a problem that happens to be a slightly changed version of a "standard" problem. An example of the first case (i.e., modify the problem) is provided by the situation in which a nonlinear objective function can be divided into "linear pieces," and then the problem is solved by the regular simplex method.† Another example of the first case is provided by the *trans-shipment problem*.‡ The trans-shipment problem differs from the standard transportation problem in one important respect. In the standard transportation problem, shipments among destinations, or among origins, are not permitted.

---

* *Goal programming*, for example, can handle the existence of multiple goals and still use the simplex method. See Lee [1972] and Hughes and Grawoig [1973, Chapter 15].
　† See Section 7.4 of Loomba and Turban [1974].
　‡ See Section 5.2.13 of Loomba and Turban [1974].

However, in many real situations it might be necessary to transport goods among origins, or destinations. This permits *any* origin or destination to serve as an intermediate source, and the problem is then referred to as the trans-shipment problem. The trans-shipment problem can be modified to obtain an equivalent standard transportation problem, and the transportation model can then be applied to solve the original problem.

An example of the case in which a slightly modified version of a standard problem is solved by a modified linear programming model is provided by integer programming. The reader will recall from Chapter 11 that if, on a linear programming problem, we impose the additional constraint that one or more of the variables must assume only integer values, we obtain an integer programming problem. Such a problem can be solved by the branch-and-bound method (Chapter 11) or by a slightly modified version of the simplex method.* Similarly, goal programming, which can handle a problem with multiple goals, is also a simplex-based method.

The *fourth* level of the application phase considers the specific manner of deriving the model solution and is, therefore, related both to the three levels of application described earlier and to the implementation phase to be presented in Section 12.3. There are essentially two ways of making the necessary computations and deriving the model solution: (1) manual solution, and (2) computer solution. The model solution can be derived manually if the problem size is relatively small. Deriving the solution manually is also quite helpful in the initial stages of learning the mechanics and rationale of linear programming. It is for this reason that we chose the manual option with respect to this introductory book on linear programming. However, once the basic learning stage has been completed, the reader must obtain a working familiarity with using a computer system as the means to derive solutions to linear programming problems.

Most real-world linear programming problems are of large size and simply cannot be solved manually. For example, it is not uncommon to find practical problems that, when expressed in the linear programming format, contain hundreds of constraints (rows) and hundreds or even thousand of activities (columns). For such large problems, the availability of, or access to, a high-speed computer is a practical necessity.

12.2.1 What Is a Computer System?

A computer is nothing more than a machine that can perform arithmetic operations and make logical choices with high speed and precision.† A

---

* For various methods of solving an integer programming problem, see Chapter 6 of Loomba and Turban [1974].

† The discussion here pertains to a *digital* computer, in which the data are represented by discrete sets of digits and are based on the concept of counting. In the *analog* computer, the data are represented in a continuous form by a measurable physical quality (based on the concept of measurement).

computer system essentially consists of five basic components:* (1) hardware, (2) software, (3) programs (4) procedures, and (5) personnel. The term *hardware* refers to the equipment part of the computer system, the main computer and its auxiliary parts. The term *software* refers to a program, code, or a set of instructions that must be fed to the computer to execute computing tasks. A *program* is a set of instructions, provided by the computer manufacturer, obtained from an outside vendor of software packages, or designed by the user to perform specialized task. A program is written in one of the computer languages, such as FORTRAN, BASIC, and COBOL.† For example, it is a simple task to write a FORTRAN program that will systematically execute all the steps of the simplex method. In practice, it is not necessary to write one's own program, because several standard software packages are commercially available. (Some of the important codes are listed in Section 12.2.4.)

A computer *procedure* refers to instructions regarding the procurement and preparation of data, operation of the computer, and how to deal with operational contingencies. The *personnel* necessary to a computer system consist of systems analysts, programmers, and computer operators.‡ The total computer system performs the essential functions of preparing the data, inputting the data (along with a program or set of instructions) to the computer in a machine-readable form, processing of the data by the main unit, storage of the data in secondary or auxiliary units, obtaining the output from the computer in a specified format, and controlling or monitoring both the input and the output. The basic organization of a computer is shown in Figure 12.1.

12.2.2   Time Sharing

In this section we explain two modes of computer processing and give a very brief description of what is known as a time-sharing system. This knowledge will help us in understanding the solution presented in Section 12.2.3 of a linear programming problem that was produced by using such a time-sharing system.

There are essentially two modes of computer processing: (1) *batch processing*, and (2) *real-time processing*. In batch processing, the data are accumulated over a period of time and then processed. The processing of a payroll by using the data contained in keypunched cards is an example of batch processing. In real-time processing, the data are not accumulated over a period of time; they are processed as soon as they are received. Examples of

---

* See Davis [1971, pp. 13–29].

† For a discussion of the computer languages, see Chapter 12 of Davis [1971]. A computer language provides the means to communicate with the computer. In practice, a set of routines, called a *compiler*, translates a program written in, say, FORTRAN, to a machine-readable form.

‡ For job descriptions, see Davis [1971, p. 16].

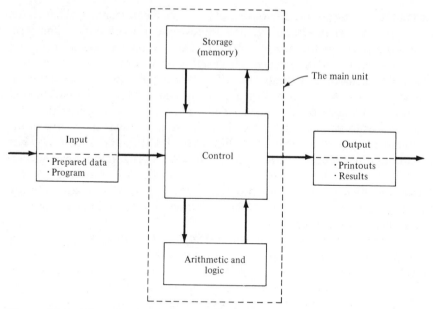

Figure 12.1 Organization of a Computer

real-time processing include airline reservation systems. No waiting is required of a customer who is seeking information regarding the availability
of space on a given flight.

A *time-sharing system* exists when many users can *simultaneously* use
a single computer system. The name "time-sharing" is derived from the
fact that multiple users can directly and concurrently "share" the "time" of
a large computer system. Use of time-sharing systems, which are based on the
concept of economy of scale, is growing rapidly. In a typical time-sharing
system, the central computer system is located at a central location while
the users are at different and often distant places. The central computer
system consists essentially of a set of computers that perform the tasks of
sequencing incoming requests, performing the necessary calculations
according to a set of instructions, and then sequencing the outgoing re-
sponses. The users of the time-sharing system communicate with the central
computer from remote terminals. The terminals, located at various geo-
graphical locations (the terminal can be in an office, a home, a library, or a
classroom), are connected with the central computer system through specially
designed telephone lines. The user sends his request to the central computer
by typing his data and indicating the program on an input device such as a
teletype.* The incoming request is handled by a communications computer
(at the central location) and, in sequence, is fed into the processing computer
(also located at the central location). Each user is allocated a unit of time for

---

* A teletype is a device that combines the features of a telephone and a typewriter.

processing his request. If the job is not completed within the allocated time, the problem is put back in computer storage until the next turn for that particular user. This pattern is repeated until the job is done and the response is sequenced back to the user through the communications center. The response is produced in the form of a printout at the remote terminal. The user can interact with the system by asking specific questions (e.g., questions on sensitivity analysis, discussed in Chapter 8). All the activities in a time-sharing system are controlled by a supervisory or executive program stored in the computer memory. And, all this happens at such a fast pace that each user feels as if he had an exclusive use of the entire computer system.

12.2.3   Computer Solution: An Illustrative Example

Having described the basic components of a computer system, the functions and interrelationships of these components, and the various modes of access to a computer resource, we are in a position to better appreciate the role of computers in linear programming. In this section we shall discuss and illustrate a computer-produced solution of the linear programming problem of Table 3.2. When stated mathematically, the problem of Table 3.2 becomes:

Maximize $\quad\quad\quad\quad\quad 23X + 32Y$

subject to the constraints

$$10X + 6Y \leqslant 2{,}500$$
$$5X + 10Y \leqslant 2{,}000 \quad\quad\quad (12.1)$$
$$1X + 2Y \leqslant 500$$

and

$$X, \quad\quad Y \geqslant \quad 0$$

A linear programming code, or system, or package, can range from a small, simple program to a very large, complex program. A simple program can be limited in scope to nothing more than producing the intermediate solution stages and the final optimal solution. A large, complex program can be designed to perform a variety of tasks, such as (1) finding the optimal solution; (2) identifying the basic and nonbasic variables; (3) identifying the shadow prices of resources; (4) conducting sensitivity analysis in terms of changes in C_j, a_{ij}, right-hand-side constraints, addition or deletion of a product, and addition or deletion of a resource; (5) arranging the output in various formats; and (6) producing routines that can update data, aggregate problems, retrieve special input information, give information on lower and upper bounds of different parameters, and so on.* In this illustration we shall restrict ourselves to discussing a simple program whose output is shown in Table 12.1.

* See the various codes listed in Loomba and Turban [1974, pp. 170–71].

Table 12.1. Computer-Produced Solution to the Problem Stated in (12.1)

---

1. s
2. PROJECT NUMBER, ID? bba00101, bba00101
3. ACTIVE SYSTEMS: WYLBUR, CALLOS
4. SYSTEM? CALLOS
5. ON AT 19:01   01/21/75 CUNY LINE 52
6. USER NUMBER, PASSWORD?
7. BBA))!
8. USER NUMBER, PASSWORD?
9. BBA001, BBA001
10. READY
11. RUN **LINPRO
12. LINPRO   19:02   01/21/75
13. LINEAR PROGRAMMING WRITTEN BY MOTOKI SHIRASUKA OF BARUCH COLLEGE
14. OF THE CITY UNIVERSITY OF NEW YORK IN JULY, 1971
15. TYPE '0' FOR THE SOLUTION ONLY, '1' FOR THE BASIS ONLY OR '2'
16. FOR THE OUTPUT IN FULL. WHICH DO YOU WANT?
17. ?2
18. WHAT ARE M (NUMBER OF ROWS (CONSTRAINTS)) AND N (NUMBER OF VARIABLES)
19. OF THE DATA MATRIX?
20. ?3, 2
21. HOW MANY 'LESS THANS,' 'EQUALS TO,' AND 'GREATER THANS' DO YOU HAVE
22. IN YOUR CONSTRAINTS?
23. ?3,0,0
24. ENTER THE COEFFICIENTS OF THE VARIABLES IN EACH RESTRICTION, ONE LINE
25. AT A TIME, SEPARATING THE COEFFICIENTS BY COMMAS
26. ?10,6
27. ?5,10
28. ?1,2
29. ENTER THE CONSTRAINTS THAT MAKE UP THE RIGHT SIDE OF THE RESTRICTIONS,
30. SEPARATING THE VALUES BY COMMAS
31. ?2500,2000,500
32. ENTER THE COEFFICIENTS OF THE LINEAR OBJECTIVE FUNCTION

Table 12.1 (*Continued*)

---

33. ?23,32
34. YOUR VARIABLES   1 THROUGH 2
35. SLACK VARIABLES   3 THROUGH 5
36. TABLEAU AFTER   0 ITERATIONS:

10.000	6.000	1.000	0.0	0.0	2500.000
5.000	10.000	0.0	1.000	0.0	2000.000
1.000	2.000	0.0	0.0	1.000	500.000
-23.000	-32.000	0.0	0.0	0.0	0.0

BASIS BEFORE ITERATION   1

VARIABLE	VALUE
3.000	2500.000
4.000	2000.000
5.000	500.000

OBJECTIVE FUNCTION VALUE   0.0
BASIS BEFORE ITERATION   2

VARIABLE	VALUE
3.000	1300.000
2.000	200.000
5.000	100.000

OBJECTIVE FUNCTION VALUE   6400.000
ANSWERS:

VARIABLE	VALUE
1.000	185.714
2.000	107.143
5.000	100.000

OBJECTIVE FUNCTION VALUE   7699.996
DUAL VARIABLES:

COLUMN	VALUE
3	1.000
4	2.600
5	0.0

TABLEAU AFTER   2 ITERATIONS:

1.000	0.0	0.143	-0.086	0.0	185.714
0.0	1.000	-0.071	0.143	0.0	107.143
0.0	0.0	0.0	-0.200	1.000	100.000
0.0	0.0	1.000	2.600	0.0	7699.996

STOP
TIME 0.8 SECS.

The printout shown in Table 12.1 is the solution to our product-mix problem of Table 3.2. The computer program employed to solve the product-mix problem was developed by Motoki Shirasuka of Baruch College. As the reader can verify in Table 12.1, it took only 0.8 second to solve our product-mix problem. The cost of running the program was only 33 cents. Let us discuss the output shown in Table 12.1 by relating it to the problem data given in (12.1). The discussion is summarized in Table 12.2.

Table 12.2

Line Number	Input Provided by the User	Question or Output of the Computer
1	The letter "s" signifies the starting point of the user's request.	
2		The computer responds by asking the user for his project number and ID (identification).
2	The user provides two numbers that identify his cost-allocation number and project security number.	
3–4		The computer lists the two active systems in its memory and asks the user to name the desired system.
4	The user chooses the CALLOS system.	
5		The computer says that the system is on. It provides the time, date, and university line number for computer usage.
6		The computer asks for the user number (cost allocation) and the password (project security).
7	The user gives a number that is *not* correct.	
8		The computer, having recognized the wrong input in line 7, again asks for the user number and the password.
9	This time, the user provides the correct input.	
10		The computer gives the signal that it is READY to accept the user's request.
11	The user asks for the specific program entitled LINPRO.	
12–14		The computer says it is ready with the program, gives the time and the date, and identifies the author of the program.
15–16		The computer provides the code and asks for the details of the user's request.
17	The user requests for the FULL OUTPUT (i.e., the initial and the optimal tableaus, information regarding the *basis*, the *basic* variables, the value of the objective function after each iteration, and value of the dual variables) by inputting 2.	

Table 12.2 (*Continued*)

Line Number	Input Provided by the User	Question or Output of the Computer
18–19		The computer asks for the number of rows and column of the data matrix.
20	The user says that there are 3 rows and 2 columns in the data matrix of the problem.	
21–22		The computer asks information regarding the type of constraints.
23	The user says there are 3 "less than," 0 "equal to," and 0 "greater than" constraints.	
24–25		The computer asks for and gives directions on how to input the coefficients of the variable in each restriction or constraint.
26–28	The user provides the data on input–output coefficients (a_{ij}) in the data matrix.	
29–30		The computer asks for and provides directions on how to input the right-hand-side constants.
31	The user provides the data on the right-hand-side constants (b_i).	
32		The computer asks for the objective function coefficients.
33	The user provides the data on the objective function coefficients (C_j).	
34–35		The computer identifies the real as well as the slack variables by numbers.

The "conversation" between the computer and the user, summarized in Table 12.2, is now complete, and the computer proceeds to perform the necessary computations and provide *all* the information requested by the user. This information, shown from line 36 onward in Table 12.1, is self-explanatory. The reader should compare it with the manually derived information at different solution stages and in the optimal solution to the same problem in Chapter 7.*

* Note that the test of optimality in the computer solution requires the net evaluators to be positive in the maximization case. This is because in this program the net evaluators are $Z_j - C_j$, rather than $C_j - Z_j$ as employed in Chapter 7 and elsewhere in this book.

12.2.4 Selecting the Computer and the Code

The printout shown in Table 12.1 was produced by using the LINPRO code and a time-sharing system. The purpose here was to illustrate the idea of how a computer can be used to solve a linear programming problem. In real-life problems, the selection of the computer and the code is not an easy task. Questions relating to *speed, precision, storage size, turnaround time, cost,* and *specific needs* of the user must be considered. More than 100 linear-programming codes for a vast variety of computers are available on the market. Some are as follows:

1. IBM 1410 Basic Linear Programming System.
2. IBM System/360 Mathematical Programming System.
3. IBM System/360 MPSX.
4. IBM 1620–1311 Linear Programming System.
5. IBM 1130 Linear Programming-Mathematical Optimization Subroutine System.
6. IBM 7040/44 Linear Programming System II for the 7040/44 Operating System.
7. The Rand Corporation MFOR Program (7090).
8. CEIR, LP90 (available through SHARE as IK LP90).
9. Honeywell 800/1800 Advanced Linear Programming System (ALPS).
10. Univac 1107 Linear Programming. Univac 1108.
11. Bendix G20 Linear Programming Code One.
12. Philco LP 2000 Linear Programming.
13. Control Data Corporation 3600 Ophelie Linear Programming.
14. General Electric 225 Linear Programming.
15. Control Data Corporation 6400, Optima.

12.3 SOME ASPECTS OF THE IMPLEMENTATION PHASE

The implementation phase starts with the output produced by the computer and ends when the desired changes (as suggested by the model solution) in the operating system have been made. It is obvious that we cannot obtain the full benefits of a linear programming system,* or any other management science model or system, until and unless the model solution is *actually* implemented. It is necessary, therefore, for the manager to have some understanding of those factors that determine the degree of success in the implementation phase.

* A *linear programming system* can be viewed as one part of a fully operational management system built around the necessary computer resources and software packages.

The problem of implementation has only recently begun to receive the degree of attention that it deserves.* This is only natural, as in the early stages of development of any tool or technique the major focus of efforts is always on perfecting the technical aspects. Once the technical aspects are developed and the application potential of the model is tested in the real world, the problems of implementation begin to assume their due importance. The recent studies on the implementation problems appear to arrive at three major conclusions, which are described in the following paragraphs.

The first conclusion is that regardless of several technical and sociological obstacles the degree of implementation of linear programming and other management science models has steadily been increasing during the last three decades. This trend has been attributed to the following:

1. Success is contagious and builds on itself. Hence, when information regarding a successful implementation (with attendant cost savings) becomes known through published reports, journals, or professional societies, it spurs interest in many users who otherwise would be hesitant to undertake risks that are usually associated with untested models.
2. The level of technical education of the manager has been increasing. Hence there is better and more effective communication between the manager and the technical personnel.
3. The quality as well as the scope of the models has been increasing. Hence the manager has attempted to apply these models to a greater number of problems in diverse areas of business, industry, and government.
4. The speed and precision of the available computers have become more favorable to applications, and the cost of computers has gone down considerably.

The second conclusion is that the degree of successful implementation can be increased by considering *explicitly* the psychological and sociological factors in the design of the model. The possible conflict of goals between management scientists and the user is a case in point. The manager might feel threatened by the potential success of the model. For example, the manager might feel that the model will replace him as a decision-maker. This type of fear is, of course, not supported by evidence. The model does *not* replace the manager; it assists him in his decision making. Another source of resistance to implementation is the fear of the operating and production workers that the model might drastically alter their working conditions, and thereby reduce their work satisfaction. This fear is also unfounded. A successful implementation of management science spreads benefits at all levels in the organization and, if managed properly, can increase the quality of life of all organization members, including the production workers.

* See, for example, Duncan [1974], Hammond [1974], and Argyris [1971].

The third conclusion is that the implementation of management science models can be increased only by conducting an extensive research into the behavioral areas of interpersonal, group, and intergroup functioning.* This assertion recognizes the vital importance of linking behavioral aspects of the process of management with the technical aspects of modern management science models.† This emphasis is an inherent part of the definition of management that was developed in Chapter 1.

In a recent study, the problems of implementation have been examined by dividing the determinants of implementation into the two groups of "project particulars" and "implementation climate."‡ The project particulars refer to such factors as cost–benefit projections, technical feasibility, priority of the project, and so on. The implementation climate refers to the sociological and psychological environment in which the projects are implemented. The implementation climate determines the organizational viability of the project. A delineation of these two distinct, but intimately related sets of determinants, is an important element of any implementation effort. The manager must understand the nature, role, and impact of these and related factors that are crucial in the application and implementation of management science models.

12.4 SUMMARY

The main emphasis of this book has been on the philosophy and logic of linear programming, rather than on the mathematics and mechanics. Yet, it is extremely difficult to provide substance to the logic, especially in the area of quantitative models, without the assistance of numerical illustrations. Such an approach is particularly useful at the introductory level. That is why the entire book was built around the core of a simple problem taken from a real-life situation. Once the philosophy, logic, mechanics, and the model structure of linear programming are grasped, it is desirable that the manager familiarize himself with some of the important aspects of application and implementation.

In this chapter we have described and discussed the question of application at four different but related levels. A very brief description of a computer system was given as a prelude to illustrating a computer-produced solution of the same product-mix problem that has been solved by different methods in this book. A line-by-line interpretation of the computer printout produced by a simple computer code, LINPRO, was given. We have also provided a selected list of other linear programming codes that are being used in industry.

* Dalton [1968, Chapter 5].
† Schultz and Slevin [1975].
‡ See Turban and Meredith in Turban and Loomba [1976].

Implementation that produces measurable changes is the final test of a successful model. We have provided only the briefest discussion of this most important aspect of management science models. A summary of the conclusions drawn from some recent studies of implementation was given. This is obviously no more than scratching the surface. The reader is encouraged to further explore this matter.

Finally, and perhaps most important, it should be stated that linear programming should be viewed as no more than a logical approach to management. Further, it is only one of the vehicles, tools, or models that "signify" the future potential of management science and operations research. The full impact of, and benefits from, these models can be realized only when the manager employs an integrated approach that consciously combines the technical elements with the behavioral factors.

REFERENCES

Argyris, C. "Management Information System: The Challenge to Rationality and Emotionality." *Management Science*, **17**, No. 6 (Feb. 1971), B275–92.

Childress, R. L. *Sets, Matrices, and Linear Programming*. Englewood Cliffs, N.J.: Prentice-Hall, Inc., 1974.

Cooper, L., and D. Steinberg. *Linear Programming*. Philadelphia: W. B. Saunders Company, 1974.

Dalton, G. W., et al. *The Distribution of Authority in Formal Organizations*. Boston: Harvard University Press, 1968.

Davis, G. B. *Introduction to Electronic Computers*, 2nd ed. New York: McGraw-Hill Book Company, 1971.

Duncan, J. W. "The Researcher and the Manager: A Comparative View of the Need for Mutual Understanding." *Management Science*, **20**, No. 8 (Apr. 1974), 1157–1163.

Ghosal, A., et al. *Examples and Exercises in Operations Research*. New York: Gordon and Breach, Inc.,1975.

Hammond, J. S. "Roles of the Manager and the Management Scientist in Successful Implementation." *Sloan Management Review*, **15**, No. 2 (Winter 1974), 1–24.

Hughes, A. J., and D. Grawoig. *Linear Programming*. Reading, Mass.: Addison-Wesley Publishing Co. Inc., 1973.

Lee, S. M. *Goal Programming for Decision Analysis*. Philadelphia: Auerbach Publishers, Inc., 1972.

Loomba, N. P., and E. Turban. *Applied Programming for Management*. New York: Holt, Rinehart and Winston, Inc., 1974.

McMillan, C. *Mathematical Programming*. New York: John Wiley & Sons, Inc., 1974.

McMillan, C., and R. Gonzalez. *Systems Analysis*, 3rd ed. Homewood, Ill.: Richard D. Irwin, Inc., 1973.

Ramsing, K. D., and D. Moberg. "Motivation, Model Perversion, and the Management Science Team." *Interfaces* (Nov. 1974), 44–47.

Schultz, R. L., and D. P. Slevin (eds.). *Implementing Operations Research/Management Science: Research Findings and Implications*. New York: American Elsevier Publishing Co., Inc., 1975.

Turban, E., and N. P. Loomba. *Readings in Management Science.* Dallas, Tex.: Business Publications, Inc., 1976.

Wagner, H. M. "The ABC's of OR." *Operations Research* (Oct. 1971), 1259–1281.

Wolfe, C. S. *Linear Programming with Fortran.* Glenview, Ill.: Scott, Foresman and Company, 1973.

REVIEW QUESTIONS AND PROBLEMS

12.1. Discuss some of the most important aspects of the application phase and the implementation phase of management science models.

12.2. Describe the main components of a computer system. How do you view a linear programming system?

12.3. Describe how a time-sharing system works.

12.4. What in your opinion are the important determinants of implementation?

12.5. What type of organization climate is conducive to successful implementation?

Appendix A   The Meaning of Linearity

A.1   CONSTANTS, PARAMETERS, AND VARIABLES

In order to grasp the nature of linear programming problems, one must be familiar with such terms as variables, parameters, functions, and linear equations. An attempt will be made in the following paragraphs to give some tangible meaning to these terms.

Suppose that an industrial worker has a wage rate of $2 per hour. Then, when he works 40 hours in a particular week, his total earnings for that week are $80. During some other week he may have worked for a total of only 30 hours, in which case he would have earned $60. In any case, his total earnings for a particular week, assuming no overtime, can always be calculated as follows:

$$\text{total earnings} = 2 \times \text{the number of hours worked}$$

If we let

h = number of hours worked

T = total earnings

then

$$T = 2h$$

is the relationship between this worker's total earnings and the number of hours worked.

307

Similarly, another worker might have a wage rate of \$4 per hour; for him, the total earnings will be given by the equation $T = 4h$. Briefly, if the number of hours worked were the only criterion for wage payment, we could say that $T = Kh$, where K is a constant, for a particular worker or for a particular class of workers, to be determined in a specific situation. This type of constant, which is fixed for a specific situation, problem, or context (but can vary from one problem context to another) is called a *parameter*. This is in contrast to such *absolute* constants as *pi* (denoted by the symbol π which has a value of 3.1416) whose value remains the same in all problems, situations, and contexts. In our simple illustration we note two kinds of mathematical quantities. One type (the quantity 2 in this example) remains fixed and is the parameter. The other type, exemplified by T and h, is allowed to vary. Quantities such as h and T, since they can assume various values in a given problem, are called *variables*. Further, since the value of T is dependent on h and is determined by assigning a value to h, T is called a *dependent variable*, and h is called an *independent variable*.

A.2 THE CONCEPT OF A FUNCTION

A *function* is a rule that relates different variables. The fact that in the preceding section T was determined by h can also be represented by the notation $T = f(h)$, which reads: "Total earnings are a function of hours worked." It does *not* mean that T is equal to f times h.

The total earnings of our industrial worker were determined by only one independent variable, h. In some cases, more than one independent variable may determine the value of the dependent variable. For example, for a company that produces only one type of product, profit is determined by taking the difference between unit revenue and unit cost for that product and multiplying it by the number of units sold. This general fact can be stated with the following notation:

$$p = f(n, r, c) \tag{A.1}$$

where

$p =$ total profit (dollars)

$n =$ number of units sold (physical units)

$r =$ revenue per unit (dollars per unit)

$c =$ cost per unit (dollars per unit)

Equation (A.1) reads: "Total profit is a function of n, r, and c."

This functional notation is meant to give the generalized idea that certain variables are somehow related. An explicit statement of equality among the variables will particularize the relationship. For example, if $r = \$4$ per unit and $c = \$3$ per unit, the general functional relationship given by Equation (A.1) will become: $p = n(4 - 3)$. Similarly, in our earlier example, the equation $T = 2h$ particularized the general functional relationship $T = f(h)$. In other words, a rule of correspondence between h and T was given. Thus a function is sometimes defined as a rule of correspondence among variables.

As mentioned earlier, once a function is particularized, it establishes an equality between different quantities on opposite sides of an equal sign. Therefore, an equation is essentially a statement of equality among different quantities.

A.3 DOMAIN AND RANGE

Let us go back to our original example. The worker may be employed in a factory in which overtime work is never permitted. Obviously, his weekly working hours can vary only from a minimum of zero hours to a maximum of 40 hours. In other words, h can only assume one of the following values:

$$h = 0, 1, 2, \ldots, 40$$

The possible values that the independent variable h can assume are said to comprise the *range* of the independent variable and the *domain* of the function.

Corresponding to each of the possible values of h, the dependent variable T takes the following values:

$$T = 0, 2, 4, \ldots, 80$$

The possible values that the dependent variable T may take are said to comprise the range of the dependent variable or the range of the function.

A.4 SINGLE-VALUED VERSUS MULTIPLE-VALUED FUNCTIONS

In our example, for every possible value of the independent variable h, the dependent variable T obviously takes one and only one value. A function such as this ($T = Kh$) is said to be a single-valued function of h. On the other hand, in the equation of a circle with center at the origin ($x^2 + y^2 = r^2$), for a given radius r (say 2 feet), we note a functional relation of the form $y = f(x, r)$, which is not single-valued. In particular, here,

$$y^2 = r^2 - x^2 = 4 - x^2$$

or

$$y = \pm\sqrt{4 - x^2}$$

In other words, for a particular value of x, the variable y assumes two values. This type of function, then, is a multiple-valued function of x. When working with multiple-valued functions, the analyst must choose only the solution(s) applicable to the problem at hand. In linear programming problems, we deal with single-valued linear functions only.

A.5 LINEAR EQUATIONS

A *linear equation* is a relationship between quantities in which, when reduced to its simplest form, the sum of the exponents of the variables in any one term adds up to 1. For example, $Y = 4 + 2X$ is a linear equation that relates the variables X and Y. When we say that the relationship between, say, two variables X and Y is linear, we mean that additions of the same magnitude to the one have a constant effect on the other. If we assume X to be the independent and Y the dependent variable, the value of Y is determined by the value assigned to or taken by X. Thus we say that Y is a function of X. The functional statement represented by the notation $Y = f(X)$ states that Y is a function of X but does not give any information about the exact relationship between X and Y. On the other hand, $Y = f(X) = 4 + 2X$, which is a linear relationship, is a complete statement of the relationship between X and Y.

Further discussion of linear equations is presented in the following examples.

EXAMPLE A.1

The equation $X = 5$ is a linear equation. Notice that X is the only variable in this equation, and its exponent is 1. The relationship $X = 5$, if considered in only one direction, represents a point on the X axis (see Figure A.1). This point can be plotted by moving 5 units to the right of the origin O. The same equation ($X = 5$), if considered in a two-dimensional space, becomes a straight line. This straight line is obtained by first plotting a point that is 5 units to the right of the origin O and then drawing a line

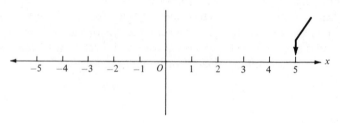

Figure A.1

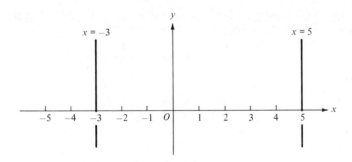

Figure A.2

through this point parallel to the Y axis (see Figure A.2). Similarly, $X = -3$ is another linear equation that can be plotted either as a point, by moving 3 units to the left of the origin O, or as a line parallel to the Y axis passing through the point $X = -3$.

EXAMPLE A.2

Now consider a linear equation of the form $Y = a + bX$, which is a relationship in two dimensions, horizontal and vertical. Traditionally, Y is plotted in the vertical direction, and X, as before, is plotted in the horizontal direction (see Figure A.3). In this case, a and b are parameters

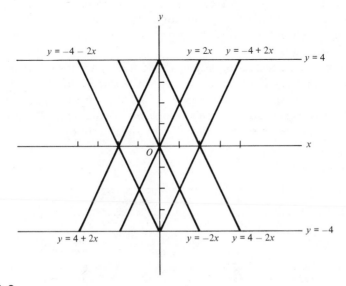

Figure A.3

(constants) whose values are to be assigned. Let us examine the following equations of the general form $Y = a + bX$:

$$Y = 4 + 2X \tag{A.2}$$

$$Y = -4 + 2X \tag{A.3}$$

$$Y = 4 - 2X \tag{A.4}$$

$$Y = -4 - 2X \tag{A.5}$$

$$Y = 2X \tag{A.6}$$

$$Y = -2X \tag{A.7}$$

$$Y = 4 \tag{A.8}$$

$$Y = -4 \tag{A.9}$$

These equations are plotted in Figure A.3. In each case the graph of the relationship, when plotted on the XY plane, becomes a straight line. Further, in each case the variable Y takes on a certain value when the variable X is zero. In Equation (A.2), for example, when $X = 0$, $Y = 4$. In Equation (A.3), on the other hand, when $X = 0$, $Y = -4$. Vertical distances such as 4 and -4, obtained by letting X equal zero in equations of the form $Y = a + bX$, are called Y *intercepts*. In other words, a is the intercept in the linear equation $Y = a + bx$.

Observe that the coefficient of X in Equations (A.2) and (A.3) is the quantity $+2$. This coefficient of the independent variable X is called the *slope* of the equation. Briefly, the slope represents a rate. In Equation (A.2) the slope $+2$ means that, for every unit of change in the independent variable X, the dependent variable Y changes by 2 units in the *same* direction. In Equation (A.4), on the other hand, since the slope is -2, the dependent variable Y also changes by 2 units for every unit of change in the variable X, but its direction of change is *opposite* the direction of change of the independent variable X.

Thus, if the coefficient of X in an equation of the form $Y = a + bX$ is a positive quantity, the equation has a positive slope, and changes in the variables take place in the same direction. By the same token, a negative coefficient of X indicates a negative slope, and changes in the dependent and independent variables take place in opposite directions. Table A.1 gives the

Table A.1

X	$Y = 4 + 2X$	$Y = -4 + 2X$	$Y = 4 - 2X$	$Y = -4 - 2X$	$Y = 2X$	$Y = -2X$
0	4	-4	4	-4	0	0
1	6	-2	2	-6	2	-2
2	8	0	0	-8	4	-4
3	10	2	-2	-10	6	-6
4	12	4	-4	-12	8	-8

changes in the dependent variable Y, corresponding to changes in the independent variable X, for Equations (A.2) through (A.7).

The relationship $Y = a + bX$ is called the *slope–intercept form* of a linear equation; a is the Y *intercept*, and b is the *slope*.

EXAMPLE A.3

Other types of linear equations are those in which more than two variables are related. For example, if we assume that the problem data of Table 1.1 are such that the optimal program requires full utilization of the resources of each department, we can write the following equations:

$$10X + 6Y + 2Z = 2{,}500 \quad \text{cutting department}$$
$$5X + 10Y + 5Z = 2{,}000 \quad \text{folding department}$$
$$1X + 2Y + 2Z = 500 \quad \text{packaging department}$$

Each of these equations is a three-dimensional linear equation. Any three-dimensional linear equation, if graphed, will yield a plane. Notice that in each of the above three equations the sum of the exponents of the variables in any one term is 1.

We have thus far discussed linear relationships in spaces involving not more than three dimensions. The idea of linear relationships can of course be extended to more than three dimensions, although actual graphing is not possible in such cases. We can make these general statements in connection with linear equations:

1. A linear equation involving two variables gives rise to a *straight line*.
2. A linear equation involving three variables gives rise to a *plane*.
3. A linear equation involving four or more variables gives rise to a *hyperplane*.

A.6 TESTS FOR LINEARITY OF A FUNCTION

In linear programming problems the objective function must be a linear function. In general, any first-degree polynomial such as

$$f(X) = a_1 X_1 + a_2 X_2 + \cdots + a_n X_n$$

is a linear function involving n dimensions. The mathematical test for the linearity of a given function $f(X)$ is the satisfaction of the following two conditions.

CONDITION 1 (Proportionality).

Multiplying each variable in the linear function by a constant k gives the same result as multiplying the functional value by k:

$$f(kX) = kf(X)$$

Suppose that $f(X)$ represents a profit function of the form

$$\text{profit} = f(X) = 23X_1 + 32X_2 + 18X_3$$

where the values of X_1, X_2, and X_3 represent a given product mix. Then condition 1 says that a k-fold increase in the current output (in exactly the same proportion) will result in increasing the current profit k times.

CONDITION 2 (Additivity).

If there are two separate programs for a given problem whose profit functions are represented by $f(X_1)$ and $f(X_2)$, these values are additive:

$$f(X_1) + f(X_2) = f(X_1 + X_2)$$

This condition simply means that the sum of the effectiveness (profit) of the separate programs equals the effectiveness of the joint program.

Linear programming is a linear model. A linear model is one in which the concept of "superposition" holds. In a linear system the response to every disturbance runs its course independently of preceding or succeeding inputs to the system; the total result is no more nor less than the sum of the separate components of system response.

Appendix B   A System of Linear Equations That Has a Unique Solution

Solve for x_1, x_2, and x_3 in the following set of linear equations:

$$x_1 \qquad\;\; + 2x_3 = 4$$
$$x_1 + \;\, x_2 \qquad\; = 5$$
$$4x_1 + 3x_2 + \;\, x_3 = 2$$

These equations can be transformed into matrix notation as follows:

$$\mathbf{AX} = \mathbf{P} \qquad\qquad (B.1)$$

where

$$\mathbf{A} = \begin{bmatrix} 1 & 0 & 2 \\ 1 & 1 & 0 \\ 4 & 3 & 1 \end{bmatrix} \qquad \mathbf{X} = \begin{bmatrix} x_1 \\ x_2 \\ x_3 \end{bmatrix} \qquad \mathbf{P} = \begin{bmatrix} 4 \\ 5 \\ 2 \end{bmatrix}$$

The determinant of $\mathbf{A}$ is

$$|\mathbf{A}| = \begin{vmatrix} 1 & 0 & 2 \\ 1 & 1 & 0 \\ 4 & 3 & 1 \end{vmatrix} = a_{11}C_{11} + a_{12}C_{12} + a_{13}C_{13}$$

$$= 1\begin{vmatrix} 1 & 0 \\ 3 & 1 \end{vmatrix} + 0C_{12} + 2\begin{vmatrix} 1 & 1 \\ 4 & 3 \end{vmatrix}$$

$$= 1 + 0 - 2 = -1$$

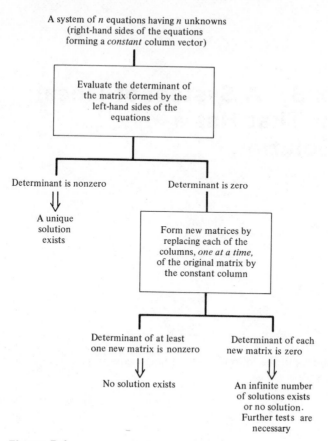

Figure B.1

Since the determinant has a nonzero value, a unique solution exists (see Figure B.1). To obtain this solution, premultiply (B.1) by the inverse of **A**. Then

$$\mathbf{A}^{-1}\mathbf{A}\mathbf{X} = \mathbf{A}^{-1}\mathbf{P}$$

or

$$\mathbf{I}\mathbf{X} = \mathbf{A}^{-1}\mathbf{P}$$

or

$$\mathbf{X} = \mathbf{A}^{-1}\mathbf{P}$$

But

$$\mathbf{A}^{-1} = \frac{\mathbf{A}_{adj}}{|\mathbf{A}|}$$

where

$$\mathbf{A_{adj}} = \begin{bmatrix} C_{11} & C_{21} & C_{31} \\ C_{12} & C_{22} & C_{32} \\ C_{13} & C_{23} & C_{33} \end{bmatrix} = \begin{bmatrix} 1 & 6 & -2 \\ -1 & -7 & 2 \\ -1 & -3 & 1 \end{bmatrix}$$

Therefore,

$$\mathbf{A}^{-1} = \frac{\mathbf{A_{adj}}}{|\mathbf{A}|} = \frac{\mathbf{A_{adj}}}{-1} = \begin{bmatrix} -1 & -6 & 2 \\ 1 & 7 & -2 \\ 1 & 3 & -1 \end{bmatrix}$$

Hence

$$\mathbf{X} = \mathbf{A}^{-1}\mathbf{P} = \begin{bmatrix} -1 & -6 & 2 \\ 1 & 7 & -2 \\ 1 & 3 & -1 \end{bmatrix} \begin{bmatrix} 4 \\ 5 \\ 2 \end{bmatrix} = \begin{bmatrix} -30 \\ 35 \\ 17 \end{bmatrix}$$

$$= \begin{bmatrix} x_1 \\ x_2 \\ x_3 \end{bmatrix} = \begin{bmatrix} -30 \\ 35 \\ 17 \end{bmatrix}$$

Thus $x_1 = -30$, $x_2 = 35$, and $x_3 = 17$. These values of x_1, x_2, and x_3 represent a unique solution to the original system of linear equations.

Appendix C A System of Linear Equations That Has No Solution

Solve for x_1, x_2, and x_3 in the following set of linear equations:

$$2x_1 + \qquad\quad 2x_3 = 4 \tag{C.1}$$

$$x_1 + x_2 \qquad = 5 \tag{C.2}$$

$$4x_1 + 3x_2 + x_3 = 2 \tag{C.3}$$

These equations can be transformed into matrix notation as follows:

$$\mathbf{AX} = \mathbf{P} \tag{C.4}$$

where

$$\mathbf{A} = \begin{bmatrix} 2 & 0 & 2 \\ 1 & 1 & 0 \\ 4 & 3 & 1 \end{bmatrix} \qquad \mathbf{X} = \begin{bmatrix} x_1 \\ x_2 \\ x_3 \end{bmatrix} \qquad \mathbf{P} = \begin{bmatrix} 4 \\ 5 \\ 2 \end{bmatrix}$$

The determinant of $\mathbf{A}$ is

$$|\mathbf{A}| = \begin{vmatrix} 2 & 0 & 2 \\ 1 & 1 & 0 \\ 4 & 3 & 1 \end{vmatrix} = a_{11}C_{11} + a_{12}C_{12} + a_{13}C_{13}$$

$$= 2\begin{vmatrix} 1 & 0 \\ 3 & 1 \end{vmatrix} + 0C_{12} + 2\begin{vmatrix} 1 & 1 \\ 4 & 3 \end{vmatrix}$$

$$= 2(1) + 0 + 2(3 - 4) = 0$$

Since the determinant is zero, no unique solution exists. In order to determine whether the system has no solution or an infinite number of solutions, we form a new matrix $\mathbf{A}'$ by replacing the third column of $\mathbf{A}$ with the constant column $\mathbf{P}$.* Thus

$$\mathbf{A}' = \begin{bmatrix} 2 & 0 & 4 \\ 1 & 1 & 5 \\ 4 & 3 & 2 \end{bmatrix}$$

The determinant of $\mathbf{A}'$ is

$$|\mathbf{A}'| = \begin{vmatrix} 2 & 0 & 4 \\ 1 & 1 & 5 \\ 4 & 3 & 2 \end{vmatrix} = a_{11}C_{11} + a_{12}C_{12} + a_{13}C_{13}$$

$$= 2\begin{vmatrix} 1 & 5 \\ 3 & 2 \end{vmatrix} + 0C_{12} + 4\begin{vmatrix} 1 & 1 \\ 4 & 3 \end{vmatrix}$$

$$= -26 - 4 = -30$$

The determinant of the new matrix $\mathbf{A}'$ is nonzero. That is, the determinant of *at least* one new matrix is nonzero. Hence the original system of equations has no solution.

To illustrate this, we first solve for x_1 and x_2 from Equations (C.1) and (C.2), in terms of x_3. From Equation (C.1),

$$x_1 = \tfrac{1}{2}(4 - 2x_3) = 2 - x_3 \tag{C.5}$$

Substituting Equation (C.5) in (C.2), we find that

$$x_2 = 5 - x_1 = 5 - (2 - x_3) = 3 + x_3 \tag{C.6}$$

Now, to solve for x_1 and x_2, in terms of x_3, from Equations (C.2) and (C.3), we proceed as follows. When we multiply Equation (C.2) by 3, we obtain

$$3x_1 + 3x_2 = 15 \tag{C.7}$$

Subtracting Equation (C.7) from (C.3) yields

$$x_1 = -13 - x_3 \tag{C.8}$$

Substituting Equation (C.8) in (C.2), we obtain

$$x_2 = 5 - x_1 = 5 - (-13 - x_3) = 18 + x_3 \tag{C.9}$$

Clearly, the values of x_1 and x_2 in Equations (C.5) and (C.6) do not agree with the values of x_1 and x_2 in Equations (C.8) and (C.9). Hence the original system of equations is inconsistent and has no solution.

* See Figure B.1.

Appendix D A System of Linear Equations That Has an Infinite Number of Solutions

Solve for x_1, x_2, and x_3 in the following set of linear equations:

$$2x_1 \quad\quad + 2x_3 = 4 \tag{D.1}$$

$$x_1 + x_2 \quad\quad = 2 \tag{D.2}$$

$$4x_1 + 3x_2 + x_3 = 8 \tag{D.3}$$

These equations can be transformed into matrix notation as follows:

$$\mathbf{AX} = \mathbf{P} \tag{D.4}$$

where

$$\mathbf{A} = \begin{bmatrix} 2 & 0 & 2 \\ 1 & 1 & 0 \\ 4 & 3 & 1 \end{bmatrix} \quad \mathbf{X} = \begin{bmatrix} x_1 \\ x_2 \\ x_3 \end{bmatrix} \quad \mathbf{P} = \begin{bmatrix} 4 \\ 2 \\ 8 \end{bmatrix}$$

The determinant of $\mathbf{A}$ is

$$|\mathbf{A}| = \begin{vmatrix} 2 & 0 & 2 \\ 1 & 1 & 0 \\ 4 & 3 & 1 \end{vmatrix} = 0$$

Since the determinant is zero, no unique solution exists. To determine whether the system has no solution or an infinite number of solutions, we form a new matrix $\mathbf{A}'$ by replacing the first column of $\mathbf{A}$ with the constant column $\mathbf{P}$.* Thus

$$\mathbf{A}' = \begin{bmatrix} 4 & 0 & 2 \\ 2 & 1 & 0 \\ 8 & 3 & 1 \end{bmatrix}$$

* See Figure B.1.

The determinant of $\mathbf{A}'$ is

$$|\mathbf{A}'| = \begin{vmatrix} 4 & 0 & 2 \\ 2 & 1 & 0 \\ 8 & 3 & 1 \end{vmatrix} = a_{11}C_{11} + a_{12}C_{12} + a_{13}C_{13}$$

$$= 4\begin{vmatrix} 1 & 0 \\ 3 & 1 \end{vmatrix} + 0C_{12} + 2\begin{vmatrix} 2 & 1 \\ 8 & 3 \end{vmatrix}$$

$$= 4(1) + 0 + 2(6 - 8) = 4 - 4 = 0$$

The determinant of $\mathbf{A}'$ is zero. Similarly, it can be shown that the matrices formed by replacing, respectively, the second and the third column of $\mathbf{A}$, by the constant column will have zero determinants. Hence the system has either an infinite number of solutions or no solution. In this case we have an infinite number of solutions.

To illustrate this, we examine the original matrix $\mathbf{A}$ for some 2×2 submatrix whose determinant is nonzero. Let us consider the submatrix $\begin{bmatrix} 2 & 0 \\ 1 & 1 \end{bmatrix}$, which is associated with x_1 and x_2 in Equations (D.1) and (D.2). We treat x_3 as a constant, transfer it to the right-hand side of Equation (D.1), and examine the following version of Equations (D.1) and (D.2):

$$2x_1 \qquad = 4 - 2x_3 \qquad\qquad \text{(D.5)}$$

$$x_1 + x_2 = 2 \qquad\qquad \text{(D.6)}$$

From Equation (D.5),

$$x_1 = 2 - x_3 \qquad\qquad \text{(D.7)}$$

Substituting Equation (D.7) in (D.6), we find that

$$x_2 = 2 - x_1 = 2 - (2 - x_3) = x_3 \qquad\qquad \text{(D.8)}$$

Now, since x_3 is treated as a constant, it can equal any number, say C. Then, from Equation (D.8),

$$x_2 = x_3 = C$$

and from Equation (D.7),

$$x_1 = 2 - x_3 = 2 - C$$

Since C can be given any value, the original system of equations has an infinite number of solutions.

Appendix E   Pivoting

E.1   INTRODUCTION

The purpose of this appendix is to illustrate how, through the process of *pivoting*, we can transform a given simplex tableau in order to produce another tableau. We first present and explain the idea of *row operations*. Next, we describe the mechanics of pivoting in terms of four specific steps. Finally, we illustrate the use of pivoting by solving a maximization linear programming problem.

E.2   ROW OPERATIONS

All of us are familiar with the *elimination method* of solving a system of linear equations. The elimination method is based on simple algebraic operations of multiplication, addition, and subtraction. It consists of making the coefficients of all but one variable zero, and then solving for that variable whose coefficient was not zero. The value of this variable is then inserted in the appropriate equation in order to solve for the remaining variables, either directly or again by the elimination technique. We shall solve the following system of equation by the elimination method.

$$10X + 6Y = 2,500 \qquad (E.1)$$

$$5X + 10Y = 2,000 \qquad (E.2)$$

Divide Equation (E.1) by 10:

$$X + \tfrac{3}{5}Y = 250 \qquad \text{(E.3)}$$

Multiply Equation (E.3) by 5:

$$5X + 3Y = 1{,}250 \qquad \text{(E.4)}$$

Subtract Equation (E.4) from (E.2):

$$7Y = 750 \qquad \text{(E.5)}$$

Divide Equation (E.5) by 7:

$$Y = \tfrac{750}{7} \qquad \text{(E.6)}$$

Substitute Equation (E.6) in (E.3):

$$X = \tfrac{1{,}300}{7}$$

The type of basic algebraic operations that we have just performed can also be performed on the rows of a matrix. When performed on the rows of a matrix, they are called *row operations*. In particular, we define three rules of row operations:

1. Any row can be multiplied by a positive or negative constant.
2. A multiple of one row can be added to another row.
3. Any two rows of a matrix can be interchanged.

Let us apply the first two rules of row operations in order to solve the system of equations given in (E.1) and (E.2). Note that our system of equations represents the first two rows of Table 3.2. It consists of a 2×2 input–output coefficient matrix on the left-hand side of the equalities, and a constant or resource vector on the right-hand side. We shall express the system as a matrix that clearly shows the input–output coefficient matrix $\begin{bmatrix} 10 & 6 \\ 5 & 10 \end{bmatrix}$ and the resource vector $\begin{bmatrix} 2{,}500 \\ 2{,}000 \end{bmatrix}$. In addition, the same series of row operations that are performed to solve the given system of equations will be applied, in successive steps, to an identity matrix. This is shown in the third column.

First Column (Algebraic Statement)	Second Column (Matrix Notation)	Third Column
$\begin{aligned}10X + 6Y &= 2{,}500 \\ 5X + 10Y &= 2{,}000\end{aligned}$	$\begin{bmatrix} 10 & 6 & \vert & 2{,}500 \\ 5 & 10 & \vert & 2{,}000 \end{bmatrix}$*	$\begin{bmatrix} 1 & 0 \\ 0 & 1 \end{bmatrix}$

* This is an example of what is known as a "partitioned" matrix. The vertical bar indicates the place of partition.

First-row operation: Divide the first row by 10:

$$X + \tfrac{3}{5}Y = 250 \qquad \begin{bmatrix} 1 & \tfrac{3}{5} & | & 250 \\ 5 & 10 & | & 2{,}000 \end{bmatrix} \qquad \begin{bmatrix} \tfrac{1}{10} & 0 \\ 0 & 1 \end{bmatrix}$$
$$5X + 10Y = 2{,}000$$

Second-row operation: Multiply the first row above by 5 and subtract this multiple from the second row. Then the system becomes:

$$X + \tfrac{3}{5}Y = 250 \qquad \begin{bmatrix} 1 & \tfrac{3}{5} & | & 250 \\ 0 & 7 & | & 750 \end{bmatrix} \qquad \begin{bmatrix} \tfrac{1}{10} & 0 \\ -\tfrac{1}{2} & 1 \end{bmatrix}$$
$$7Y = 750$$

Third-row operation: Divide the second row above by 7:

$$X + \tfrac{3}{5}Y = 250 \qquad \begin{bmatrix} 1 & \tfrac{3}{5} & | & 250 \\ 0 & 1 & | & \tfrac{750}{7} \end{bmatrix} \qquad \begin{bmatrix} \tfrac{1}{10} & 0 \\ -\tfrac{1}{14} & \tfrac{1}{7} \end{bmatrix}$$
$$Y = \tfrac{750}{7}$$

Fourth-row operation: Multiply the second row above by $\tfrac{3}{5}$ and subtract this multiple from the first row:

$$X = \tfrac{1{,}300}{7} \qquad \begin{bmatrix} 1 & 0 & | & \tfrac{1{,}300}{7} \\ 0 & 1 & | & \tfrac{750}{7} \end{bmatrix} \qquad \begin{bmatrix} \tfrac{1}{7} & -\tfrac{3}{35} \\ -\tfrac{1}{14} & \tfrac{1}{7} \end{bmatrix}$$
$$Y = \tfrac{750}{7}$$

What the matrix notation (after the fourth row operation) shows is this:

1. We have an identity matrix on the left-hand side of the partition.
2. The identity matrix, associated with the variables X and Y, shows that variables X and Y are the basic variables. Our *basis* in this case is
$$\begin{bmatrix} 10 & 6 \\ 5 & 10 \end{bmatrix}.$$
3. The values on the right-hand side of the partition are the values of the basic variables X and Y.

What we have done is this: By performing a series of row operations on a partitioned matrix (consisting of an input–output coefficient matrix and the constant vector), we produced an identity matrix and simultaneously determined the values of the basic variables. Also, by performing the same row operations, in successive steps, on an identity matrix, we have produced the inverse of the original matrix. Note that

$$\begin{bmatrix} \tfrac{1}{7} & -\tfrac{3}{35} \\ -\tfrac{1}{14} & \tfrac{1}{7} \end{bmatrix}$$

is the inverse of

$$\begin{bmatrix} 10 & 6 \\ 5 & 10 \end{bmatrix}$$

It should now be clear that the rules of transformation of the simplex method perform exactly the same function. That is, once a basis is identified, the tableau must produce an identity matrix (under the chosen basic variables), generate the values of the basic variables under the "Quantity" column, and create an inverse of the basis under a set of nonbasic variables.

E.3 PIVOTING

The above discussion was presented to introduce the reader to the process and mechanics of *pivoting*. The use of row operations to produce a simplex tableau that represents a given basis is known as *pivoting*. The purpose of pivoting is to help us proceed from a nonoptimal simplex tableau to the next tableau, which is then tested for optimality by the rules established in Chapter 7. The process of pivoting consists of the following four steps.

STEP 1. Identify the pivot column.

The pivot column is identified by the largest positive $C_j - Z_j$ (in maximization case) or largest negative $C_j - Z_j$ (in minimization case). The *pivot column* is the same as the *key column* described in Chapter 7. The pivot column identifies the incoming variable entering the basis.

STEP 2. Identify the pivot row.

We calculate replacement quantities (for each row) by dividing the numbers under the "Quantity" column by the corresponding positive (a_{ij}) under the pivot column. The row in which falls the smallest replacement quantity is the *pivot row*. The *pivot row* is the same as the *key row* described in Chapter 7. The pivot row identifies the outgoing variable leaving the basis.

STEP 3. Identify the pivot element.

The element at the intersection of the pivot column and the pivot row is the *pivot element*. The *pivot element* is the same as the key number described in Chapter 7.

STEP 4. Transform the current tableau to produce a tableau that represents the new basis.

It is in this step that the *mechanics* of pivoting are somewhat different from the rules of transformation that were employed in Chapter 7 to proceed from one simplex tableau to the next. It will be recalled that in Chapter 7 we stated two rules of transformation: one for transforming the key row, and the other for the nonkey rows. By applying these two rules of transformation, we can proceed from one tableau to the next. The same results can be produced by pivoting. The idea of pivoting is based on the concept that when we proceed from one tableau to the next, all we do is to create a new basis and that there must exist an identity matrix under the new basic variables. *Since the pivot column represents the only new incoming variable, we can produce the new identity matrix (corresponding to the new basis) if, through the use of row operations, we obtain 1 at the pivot element and 0's elsewhere in the pivot column.*

E.4 ILLUSTRATIVE EXAMPLE

We shall illustrate the mechanics of pivoting by using the same linear programming problem that was solved in Chapter 7. Let us start with Table E.1, which represents the first program to the problem stated in Table 7.3. Remember that our task is to produce, through the use of row operations, a vector $\begin{bmatrix} 0 \\ 1 \\ 0 \end{bmatrix}$ for the next tableau to replace the pivot column vector $\begin{bmatrix} 6 \\ 10 \\ 2 \end{bmatrix}$ in the current tableau. Thus we first multiply the pivot row by the reciprocal of the pivot element (or divide the pivot row by the pivot element). In our case, we multiply the pivot row by $\frac{1}{10}$ and obtain

$$200 \quad \tfrac{1}{2} \quad 1 \quad 0 \quad \tfrac{1}{10} \quad 0 \qquad\qquad (E.7)$$

This row of numbers is entered as row 2 (the Y row) in the tableau shown in Table E.2.

Our next task is to create zeros under the pivot column. The first zero (in row S_1) can be created if we multiply (E.7) by -6 and add the multiple to row S_1 of Table E.1. This will give us the following row for the next tableau (to be placed as first row):

$$1,300 \quad 7 \quad 0 \quad 1 \quad -\tfrac{3}{5} \quad 0$$

The remaining zero (in row S_3) can be created if we multiply (E.7) by -2 and add the multiple to row S_3 of Table E.1. This will give us the following row for the next tableau (to be placed as third row):

$$100 \quad 0 \quad 0 \quad 0 \quad -\tfrac{1}{5} \quad 1$$

We have now transformed the simplex tableau of Table E.1 through the process of pivoting (i.e., by using row operations). The new tableau is shown in Table E.2. Note that the second program shown in Table E.2, obtained by pivoting, is the same as the second program of Table 7.4 that was obtained by the two rules of transformation stated in Chapter 7. Since one $C_j - Z_j$ in Table E.2 is positive (maximization case), our second program is not optimal. We identify 7 as the pivot element. Our task now is to produce, through the use of row operations, a vector $\begin{bmatrix} 1 \\ 0 \\ 0 \end{bmatrix}$ for the next tableau to replace the pivot

Table E.1. First Program

Program (Basic Variables)	Profit per Unit C_b	Quantity	$C_j \to$ 23 X	32 Y	0 S_1	0 S_2	0 S_3	Replacement Quantity
S_1	0	2,500	10	6	1	0	0	$416\frac{2}{3}$
S_2	0	2,000	5	10	0	1	0	200 → outgoing variable
S_3	0	500	1	2	0	0	1	250
Z_j			0	0	0	0	0	
$C_j - Z_j$			23	32	0	0	0	

↑ incoming variable

Table E.2. Second Program

Program (Basic Variables)	Profit per Unit C_b	Quantity	$C_j \to$ 23 X	32 Y	0 S_1	0 S_2	0 S_3	Replacement Quantity
S_1	0	1,300	7	0	1	$-\frac{3}{5}$	0	$\frac{1,300}{7} \to$ outgoing variable
Y	32	200	$\frac{1}{2}$	1	0	$\frac{1}{10}$	0	400
S_3	0	100	0	0	0	$-\frac{1}{5}$	1	
Z_j			16	32	0	3.2	0	
$C_j - Z_j$			7	0	0	-3.2	0	

incoming variable

328

column vector $\begin{bmatrix} 7 \\ \frac{1}{2} \\ 0 \end{bmatrix}$ in Table E.2. This can be accomplished easily by just two row operations. First, we multiply the pivot row by $\frac{1}{7}$ to obtain

$$\frac{1,300}{7} \quad 1 \quad 0 \quad \frac{1}{7} \quad -\frac{3}{35} \quad 0 \qquad \text{(E.8)}$$

This row of numbers is entered as row 1 (the X row) in the tableau shown in Table E.3.

Table E.3. Third and Optimal Program

Program (Basic Variables)	Profit per Unit C_b	Quantity	$C_j \rightarrow 23$ X	32 Y	0 S_1	0 S_2	0 S_3
X	23	$\frac{1,300}{7}$	1	0	$\frac{1}{7}$	$-\frac{3}{35}$	0
Y	32	$\frac{750}{7}$	0	1	$-\frac{1}{14}$	$\frac{1}{7}$	0
S_3	0	100	0	0	0	$-\frac{1}{5}$	1
Z_j			23	32	1	2.6	0
$C_j - Z_j$			0	0	-1	-2.6	0

Second, we multiply (E.8) by $-\frac{1}{2}$ and add the multiple to row Y of Table E.2. This will give us the following row for the next tableau (to be placed as row Y):

$$\frac{750}{7} \quad 0 \quad 1 \quad -\frac{1}{14} \quad \frac{1}{7} \quad 0$$

Note that in the pivot column of Table E.2, we already have a zero in the third place. Hence the third row of the second tableau need not be changed, and will appear "as is" in the third row of the tableau shown in Table E.3. Note that the optimal program shown in Table E.3, and obtained by pivoting, is the same as the optimal program of Table 7.5 that was obtained by the two rules of transformation stated in Chapter 7.

Answers to Selected Problems

Chapter 3

3.4. Doors $= 15.391$
Beams $= 20.761$
Maximum profit $= \$1,807.612$

3.8. (a) $A = 500$
$B = 0$
Maximum profit $= \$100$
(c) $A = 0$
$B = 0$
Maximum profit $= 0$

3.6. $X = 5$
$Y = 0$
Minimum value $= 25$

(b) $A = 285.714$
$B = 428.571$
Maximum profit $= \$85.714$
(d) $A = 625$
$B = 0$
Maximum profit $= \$125$

Chapter 4

4.4. $X = 6$
$Y = 16$
Minimum value $= 120$

Chapter 5

5.2.
$$A = \begin{bmatrix} V_1 \\ V_2 \\ V_3 \end{bmatrix} \quad \text{where} \quad \begin{aligned} V_1 &= [2 \quad 0 \quad -1] \\ V_2 &= [1 \quad 2 \quad 4] \\ V_3 &= [0 \quad 3 \quad 2] \end{aligned}$$

$$A = [U_1 \quad U_2 \quad U_3] \quad \text{where} \quad U_1 = \begin{bmatrix} 2 \\ 1 \\ 0 \end{bmatrix} \quad U_2 = \begin{bmatrix} 0 \\ 2 \\ 3 \end{bmatrix} \quad U_3 = \begin{bmatrix} -1 \\ 4 \\ 2 \end{bmatrix}$$

$$|A| = -19$$

5.4. $\mathbf{AB} = \begin{bmatrix} 1 & 0 & 0 \\ 0 & 1 & 0 \\ 0 & 0 & 1 \end{bmatrix}$ $\mathbf{BA} = \begin{bmatrix} 1 & 0 & 0 \\ 0 & 1 & 0 \\ 0 & 0 & 1 \end{bmatrix}$

$\mathbf{A}^{-1} = \mathbf{B}$

5.6. (b) $\mathbf{P}_0 = -4\mathbf{P}_1 + \frac{11}{3}\mathbf{P}_2 - \frac{1}{3}\mathbf{P}_3$

5.8. (a) $X_1 = \frac{165}{23}$
$X_2 = \frac{290}{23}$
$X_3 = \frac{315}{23}$

(b) No solution

(c) $\begin{vmatrix} 5 & 4 & 1 \\ 0 & 2 & 4 \\ 2 & 6 & 0 \end{vmatrix} = -92$

Chapter 6

6.4. $X = 7$
$Y = 4$
Maximum value $= 59$

Chapter 7

7.6.

Variable	Amount
Rapidgrow	15.833
Fastgrow	5.833

Minimum cost $= \$185$

7.8.

Variable	Amount
A	0
B	0
C	40

Maximum profit $= \$1,900$

7.10.

Variable	Amount
X1	0
X2	4.615
X3	12.500

Minimum $Z = 73.075$

7.12.

Variable	Amount
$X1$	1.25
$X2$	0
$X3$	2

Minimum $Z = 10.5$

Chapter 8

8.4. (a) Minimize $\quad Z = 2{,}500U_1 + 2{,}000U_2 + 500U_3$

subject to the constraints

$$10U_1 + 5U_2 + 1U_3 \geqslant 23 \qquad U_1 = 1$$
$$6U_1 + 10U_2 + 2U_3 \geqslant 32 \qquad U_2 = 2$$
$$2U_1 + 5U_2 + 2U_3 \geqslant 18 \qquad U_3 = 3$$

(b) No
(c) No

8.6. (a) Doors = 15.448   (b) Doors = 40.08
       Beams = 20.69   Beams = 1.149
       Profit = \$1,806.896   Profit = \$2,211.494

Chapter 9

9.8.

Plant	Boston	Chicago	Dallas	Kansas City	Seattle
N	400	100	300		
A			400	200	
C				100	100
D					400

Plant	Boston	Chicago	Dallas	Kansas City	Seattle
N	400	100		300	
A			100		500
C			200		
D			400		

9.10. Whenever the number of occupied cells is less than $m + n - 1$, we have a degenerate transportation problem.

9.12.

Plant	A	B	C
X	100		125
Y		50	100
Z		150	225

9.14.

	Market					
Region	A	B	C	D	E	F
X	0	900	0	0	2,700	3,400
Y	2,500	0	1,050	3,200	0	0
Z	0	1,500	800	0	0	0

Cost = \$298,200

Chapter 10

10.6.

	Machine				
Job	1	2	3	4	5
A					1
B			1		
C		1			
D	1				
E				1	

Total profit = \$374

10.8.

	Territory			
Salesman	1	2	3	4
A		1		
B			1	
C	1			
D				1

Chapter 11

11.4. $X = 9$
$Y = 10$
Minimum value $= 142$

11.6. $X = 7$
$Y = 7$
$Z = 1$
Maximum value $= 224$

11.8. Newspaper ads $= 1$
Television ads $= 2$
Minimum cost $= \$17,500$

11.10. $A = 20$ units
$B = 0$
Maximum profit $= \$6,000$

Author Index

Subject Index